D0312346

CHARLES DARWIN

CHARLES DARWIN

DESTROYER OF MYTHS

ANDREW NORMAN

Skyhorse Publishing

FIRST NORTH AMERICAN EDITION 2014

First published in Great Britain in 2013 by Pen & Sword Discovery, an imprint of Pen & Sword Books Ltd

10 9 8 7 6 5 4 3 2 1

Library of Congress Cataloging-in-Publication Data is available on file.
ISBN: 978-1-62873-725-7

Printed in the United States of America

Contents

DARWIN / WEDGWOOD
FAMILY TREE
(Selected)

ERASMUS DARWIN (I) 1731-1802 = (I) MARY HOWARD 1740-1770 = (II) ELIZABETH POLE (née COLYEAR)

JOSIAH WEDGWOOD(I) 1730-1795 = SARAH WEDGWOOD 1734-1815

+4 ROBERT WARING DARWIN 1766-1848 = SUSANNAH WEDGWOOD 1765-1817

JOSIAH WEDGWOOD(II) 1769-1843 = ELIZABETH ALLEN 1764-1846 +4

MARIANNE DARWIN 1798-1858

CAROLINE SARAH DARWIN 1800-1888

SUSAN ELIZABETH DARWIN 1803-1866

ERASMUS ALVEY DARWIN (II) 1804-1881

CHARLES (ROBERT) DARWIN 1809-1882 = EMMA WEDGWOOD 1808-1896

EMILY CATHERINE DARWIN 1810-1866

WILLIAM ERASMUS DARWIN 1839-1914

ANNE ELIZABETH DARWIN 1841-1851

MARY ELEANOR DARWIN b+d1842

HENRIETTA EMMA DARWIN 1843-1930

GEORGE HOWARD DARWIN 1845-1912

ELIZABETH DARWIN 1847-1926

FRANCIS DARWIN 1848-1925

LEONARD DARWIN 1850-1943

HORACE DARWIN 1851-1928

CHARLES WARING DARWIN 1856-1858

Acknowledgements

AK Bell Library, Perth, Scotland; British Library; Darwin Correspondence Project, Cambridge University Library; Dundee Central Library, Local History Centre; Plymouth University, Department of Earth and Environmental Sciences, Plymouth, UK; Linnean Society of London; Natural History Museum; Perth and Kinross Council Archive; Perth Museum & Art Gallery; Poole Postgraduate Library; Poole Central Library; Royal Society, London; School of Geography, Scientific Manuscripts Collections, Department of Manuscripts & University Archives, University Library, Cambridge; South Place Ethical Society, Conway Hall, London; Special Collections, Centre for Research Collections, Edinburgh University Library; United Benefice of Cudham and Downe; Welsh School of Pharmacy, Cardiff.

Paul Adair; Rupert Baker, Professor Simon Baron-Cohen; Yvonne Bell; Catherine Broad; Claire Button; Professor Anthony K. Campbell; Anne Carroll; Elaine Charwat; Rosemary Clarkson; Mary Clayton; Professor Barbara Doughty; Hugh Dower; Gabriel Dragffy; Nicholas Dragffy; Stuart Hannabuss; Ellen King; Rhona Morrison; Barbara Pierce; Dr Toni Soriano Arandes; Adam J. Perkins; Gregory D. Price; the Reverend Cliff Reed; Nigel J. Savery; Angela Stone; Deirdre Sweeney; John Watson.

I am especially grateful to my beloved wife, Rachel, for all her help and encouragement.

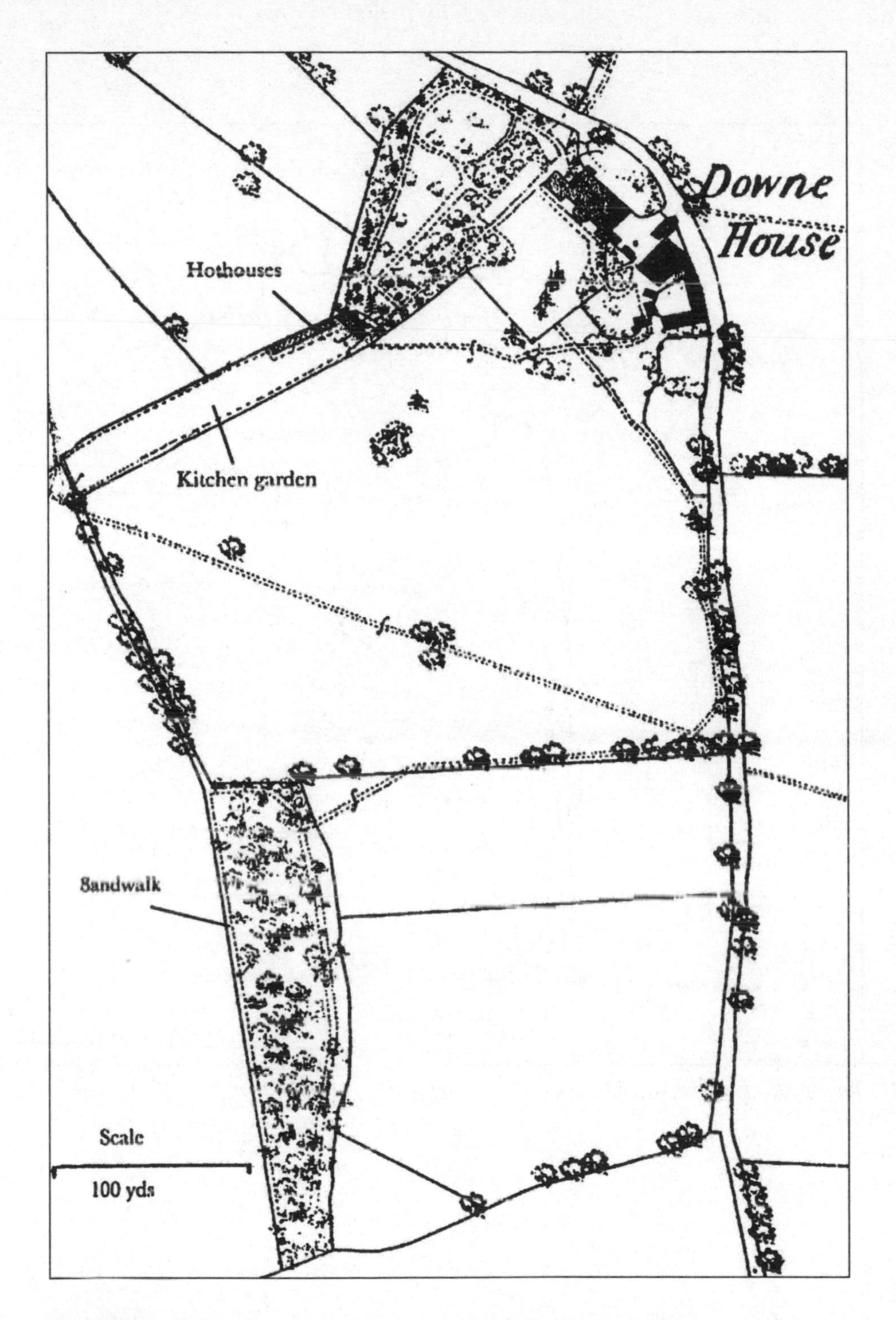

Plan of Down House and its grounds.

Places visited by HMS Beagle, *on her voyage 1831–36.*

Geographical distribution of Chagas disease vectors in Latin America. (Source: PAHO/WHO, Programme on Communicable Diseases)

Preface

Charles Darwin did not deliberately set out to be the 'destroyer of mythical beliefs', some of which, in his early days as a young Christian, he had previously espoused. He was a modest man who liked to avoid controversy of any kind, yet paradoxically, he was to be the cause of the greatest controversy in the history of the world! Neither did he quickly come to his conclusions about the origin and evolution of all life on Earth, for just as the living organisms to which his theory applied had evolved over millions of years, so his thinking evolved as his own life progressed.

Darwin was the scientific equivalent of 'Mr Valiant-for-truth', a character in writer and preacher the Reverend John Bunyan's *Pilgrim's Progress* – a religious allegory, representing the Christian journey from a sinful condition to redemption. Like 'Mr Valiant-for-truth' Darwin was to endure great hardships – 'persecution' might not be too strong a word on account of his beliefs. However, Darwin's journey was not a pilgrimage to the 'Celestial City', but one of scientific discovery.

When, in late December 1831, Darwin embarked on HMS *Beagle*, bound for the southern hemisphere, he could not have imagined that the experience would lead him to formulate a theory which would totally revolutionize the way in which man viewed the natural world. And yet, although Darwin's theory explains so much, it leaves many questions unanswered. Some relate to Darwin himself: in particular the nature of the chronic illness which plagued him all his adult life. Others relate to such questions as why did the dinosaurs become extinct; was it possible to resolve the apparent incompatibility of 'Darwinism' and science; to what extent was Darwinism a factor in the Nazi 'Holocaust'? Finally, come questions which penetrate to the very heart of what it means to be a human being.

Chapter 1

Charles Darwin: A Child Is Born

Charles Robert Darwin was born on 12 February 1809 at 'The Mount' (a mansion built by his father Dr Robert Darwin, in 1798) in Shrewsbury, the county town of Shropshire. Constructed of red brick in the late Georgian style, 'The Mount' reflected the fact that its owner was a man of substance. It was subsequently described as containing:

> Dining Room, Drawing Room, Morning Room opening into Conservatory, Library, Fourteen Bedrooms with suitable Dressing Rooms, Kitchens and all usual offices, ample Cellaring, very extensive Stabling, Coach Houses, &c., Conservatories, Fernery, Forcing Frames, extensive walled Garden, Pleasure Grounds, and adjoining piece of Land ..., and standing in an elevated position on the Banks of the River Severn, commanding [views of] extensive and beautiful scenery. The property also included a Gardener's House with Garden attached, Coach-house, Stable, &c.

The luxuriousness of life at 'The Mount' is further indicated by the presence of a parterre, a summerhouse, and an ice house (a building, typically one situated partly or wholly underground, in which food was preserved by storing it in ice).[1]

Darwin had an older brother, Erasmus Alvey (born 1804), and four sisters: Marianne (born 1798), Caroline Sarah (born 1800), Susan Elizabeth (born 1803), and (Emily) Catherine (born 1810). During his boyhood his siblings called him 'Bobby', or alternatively, 'Charley', and subsequent correspondence reveals that a strong and loving bond existed between them all.

His father Robert

Darwin's father, Robert Waring Darwin, was born on 30 May 1766. He attended Leiden University in the Netherlands where, following in the footsteps of his father, Erasmus, he qualified as a doctor.[2] He subsequently established a 'very large [medical] practice' in Shrewsbury.[3] In 1788, following the publication in the scientific journal

Philosophical Transactions (*of the Royal Society*) of a learned paper entitled 'Ocular Spectra', Robert was elected a Fellow of the Royal Society.

On 18 April 1796, Robert married Susannah, daughter of his father Erasmus's late friend, the famous Staffordshire potter Josiah Wedgwood (I).[4]

At 'The Mount', Robert 'took a great pleasure in his garden, planting it with ornamental trees and shrubs, and being especially successful in fruit-trees … .'[5] Darwin described his father as 'the kindest man I ever knew … .'[6]

His mother Susannah Darwin, née Wedgwood

Susannah came from a family whose religious persuasion was Unitarian (see below). She attended the Reverend George Case's Unitarian chapel (situated in Shrewsbury's High Street), as did Darwin and his siblings.[7] Said Darwin,

> My mother died in July 1817, when I was a little over eight years old, and it is odd that I can remember hardly anything about her except her death-bed, her black velvet gown, and her curiously constructed work-table.[8]

Dr Case and Shrewsbury Unitarian School

Apart from his ministerial duties, the Reverend Case also ran a day school in Shrewsbury that Darwin attended, the latter describing himself at this time as a collector of

> all sorts of things: shells [of marine molluscs], seals [presumably for stamping designs on documents], franks [presumably for stamping letters with official marks to record payment of postage], coins, and minerals.[9]

Darwin also reveals that he was 'interested at this early age in the variability of plants [i.e. in respect of their colours]'.[10] He made minute observations in regard to the variability of species – both plants and animals – and as a result of the deductions which he made, based upon such observations, he would one day become famous throughout the world.

Darwin confessed that, as a child, he had a penchant for telling white lies and for performing practical jokes.

> I once gathered much valuable fruit from my father's trees and hid it in the shrubbery and then ran in breathless haste to spread the news that I had discovered a hoard of stolen fruit.
>
> [However] father wisely treated this tendency not by making crimes of the fibs, but by making light of the discoveries.[11]

Darwin also described how, on one occasion, he 'acted cruelly for I beat a puppy …'. He then proceeds to analyse his emotions. Yes, he had enjoyed his 'sense of power' over the puppy, but nonetheless, 'this act lay heavily on my conscience, as is shown by my remembering the exact spot where the crime was committed'.[12]

Chapter 2

Religion: Unitarianism

Unitarianism is a so-called 'non-conformist' or 'dissenting' religion, which originated in Transylvania – now part of Romania – in the sixteenth century, and a Unitarian is defined as a person, especially a Christian, who asserts the unity of God and rejects the doctrine of the Trinity (the three persons of the Christian Godhead – Father, Son, and Holy Spirit).[1]

In England, Unitarian ideas were first expounded by John Biddle (1615–62), graduate and tutor of Magdalen Hall, Oxford and in 1774 Britain's first Unitarian congregation was established by the former Anglican clergyman Theophilus Lindsey in Essex Street, Strand, London. (Anglican – relating to or denoting the Church of England, or any Church in communion with it.)[2]

As members of a 'dissenting' religion Unitarians were liable to be ostracized for their beliefs and persecuted, and it was not until 1813 (with the repeal of certain clauses of the Toleration Act) that Unitarianism finally became a legalized form of worship.

To Unitarians Jesus Christ is not God (i.e. part of the Trinity) but rather 'a man, unequivocally human [who was] conceived and born in the usual human manner'.[3] God is therefore regarded by Unitarians as a 'unity', rather than a 'trinity' – hence the name 'Unitarian'. Unitarians also believe that following Christ's crucifixion, he did not descend into 'Hell'; for in Unitarianism, there is no such place. Neither is there such an entity as 'The Devil', and nor do they subscribe to the doctrine of 'Original Sin'. As for the 'Resurrection of Christ' from the dead – this is to be seen, not as a literal truth, but rather as a 'powerful myth';[4] and as for the notion of there being life after death 'most Unitarians agree that this is an area of mystery'.[5] To summarize

> Unitarians believe that freedom from prescribed creeds, dogma and confessions of faith is necessary if people are to seek and find truth for themselves. Shared values and a shared religious approach are a surer basis for unity than theological propositions. Because no human being and no human institution can have a monopoly of truth, it is safer to admit that from the outset. The Unitarian community is a community of the spirit that cherishes reason and acknowledges honest doubt; a community where the only theological test is that required by one's own conscience. Above all, Unitarians are bound by a sense of common humanity.[6]

Finally, Unitarians, who regard both the *Bible* and the Church as fallible, believe that for every individual, the seat of religious authority lies 'within oneself', and that 'all people develop their own belief system'.

Most importantly for the young Charles Darwin was the attitude of Unitarians to science, summed up by Unitarian minister and theologian James Martineau, Minister of Little Portland Street Chapel, London (1858–72) and Principal of Manchester New (Unitarian) College (1869–85), who declared that 'the architects of science have raised over us a nobler temple…'.[7]

Given the fact that, from an early age, Charles Darwin was passionately interested in the natural world and natural history (defined as the scientific study of the natural world, including animals and plants, palaeontology, and other natural phenomena which are the subject of scientific investigation) it was fortunate for him that the Unitarian environment (provided for him by his Unitarian chapel, Unitarian school, and Unitarian mother) was 'science friendly'. This situation, however, was now to change, for it was said of Darwin that 'after his early boyhood, he seems usually to have gone to church and not to Mr Case's [chapel].[8] This was a reference to Shrewsbury's Anglican Church of St Chad, where Darwin had been baptized in November 1809.

Robert Darwin was a 'freethinker' – one who rejects accepted opinions, especially those concerning religious belief. (Robert's father Erasmus, had also been sceptical about religion, declaring that 'Unitarianism was a feather-bed to catch a falling Christian',[9] by which he meant, presumably, that for those who had doubts about the truth of Christianity, or found it impossible to follow its tenets, there was always Unitarianism to fall back on.)

It appears, therefore, that after the death of Darwin's Unitarian mother Susannah, Robert made the decision to transfer his son (and presumably his other children) to the established Church of England. This was probably because Robert, who had ambitions for his two sons, knew that as Unitarian 'dissenters' they would face hostility and prejudice in society, which might well prove to be a hindrance to them.

Chapter 3

Shrewsbury School and the Reverend Butler

In the summer of 1818 Darwin entered Shrewsbury School as a boarder, even though the school was 'hardly more than a mile to my home'.[1] Its headmaster was Samuel Butler, who had been appointed to the post in 1798 at the age of only twenty-four years, and who would occupy this position for another thirty-eight years.

According to the school's chronicler Basil Oldham, it was said of Butler that he

> was very much alive to the claims of moral as well as intellectual education. But in the application of the most important vehicle of it, religion, the one Shrewsbury headmaster [i.e. Butler] who became a bishop [of Lichfield in 1836] was sadly wanting. He was frankly [defeatist] in regard to the practicability of influencing boys through their religion.[2]
>
> The horror he so frequently expresses of the Evangelicals [Protestant Christians who emphasize the authority of the Bible, personal conversion, and the doctrine of salvation by faith in the Atonement],[3] and the firm distinction that he draws in his sermon on Christian Liberty, between 'a Pietist and a pious person, a Puritan and a pure person, a Religionist and a religious person', show that his fear and dislike of religious enthusiasm were almost an obsession with him, and he probably felt that if formal services and formal religious instruction were of no avail with boys, he was not prepared to try any other methods.[4]

To Darwin, Shrewsbury School was a huge disappointment. He wrote:

> Nothing could have been worse for the development of my mind than Dr Butler's school, as it was strictly classical, nothing else being taught, except a little ancient geography and history. The school, as a means of education, to me was simply a blank.

Nonetheless, said Darwin, 'I was not idle, and with the exception of versification [to turn into or express in verse],[5] generally worked conscientiously at my classics, not using cribs.'[6] And being possessed of a good memory, he

> could effect with great facility, learning forty or fifty lines of Virgil or Homer, whilst I was in morning chapel; but this exercise was utterly useless, for every verse was forgotten in forty-eight hours.
>
> The only qualities which at this period promised well for the future were, that I had strong and diversified tastes, much zeal for whatever interested me, and a keen pleasure in understanding any complex subject or thing.[7]
>
> I was taught Euclid by a private tutor, and I distinctly remember the intense satisfaction which the clear geometrical proofs gave me. I remember with equal distinctness the delight which my uncle [Samuel Tertius Darwin] gave me by explaining the principle of the vernier [a small movable graduated scale for obtaining fractional parts of subdivisions on a fixed main scale][8] of a barometer.[9]

Darwin tried his best to make up for these perceived deficiencies in his education in his spare time. For example, he described reading, for pleasure, the plays of Shakespeare and the poetry of Thomson, Byron and Scott.

> Early in my school-days a boy had a copy of *The Wonders of the World*, which I often read, and disputed with other boys about the veracity of some of the statements; and I believe that this book first gave me a wish to travel in remote countries … .

Here, then, is an early clue as to the character of the young Darwin – someone who was always ready and willing to challenge the opinions of others.

But it was the natural world which fascinated him most.

> I must have observed insects with some little care, for when ten years old I went for three weeks to Plas Edwards on the sea-coast in Wales. I was very much interested and surprised at seeing a large black and scarlet Hemipterous insect, many moths and a Cicindela [brightly-coloured beetle], which are not found in Shropshire.[10]
>
> From reading White's *Selbourne* [a reference to the Reverend Gilbert White, clergyman and naturalist, whose *Natural History & Antiquities of Selborne* was published in 1879], I took much pleasure in watching the habits of birds, and even made notes on the subject.[11]

Darwin also confessed to 'collecting minerals with much zeal, but quite unscientifically …'.[12] And he 'became passionately fond of shooting'[13] (i.e. game), and enjoyed angling.

In February 1822 Erasmus (II), Darwin's elder brother, was admitted to Christ's College, Cambridge. That November, writing from Cambridge, he suggested to Darwin that their 'lab' at 'The Mount' might be improved by having 'some more shelves fixed up'. This was a reference by Erasmus to a 'fair laboratory with proper apparatus' which he had created 'in the tool-house in the garden'. Here, said Darwin, 'I was allowed to aid him as a servant in most of his experiments. He made all the gases and many compounds, and I read with care several books on chemistry … .[14]'

Furthermore, Erasmus informed Darwin that he had

> ordered a small goniometer [an instrument for the precise measurement of angles, especially one used to measure the angles between the faces of crystals[15]] so that we shall be able to separate the different [crystals] in your cab [presumably cabinet]: I have not yet procured any of the minerals you mentioned. [However] I have bought a book which will be very useful. There are directions for finding out the names of minerals &c. &c. & the rules are not very difficult. I am attending Professor Cummings's [James Cumming, Professor of Chemistry at Cambridge University] lectures on chemistry which are very entertaining. I have written all his experiments down as far as we have [proceeded] which we shall be able to try over again.[16]

A thirst for knowledge and a voracious appetite for learning were other characteristics of the young Darwin. What joy the brothers must have had together, in concocting and performing their experiments! But on the downside, said Darwin

> The fact that we [he and Erasmus] worked at chemistry somehow got known at the school, and as it was an unprecedented fact, I was nicknamed 'Gas'. I was also once publicly rebuked by the headmaster, Dr Butler for thus wasting my time on such useless subjects; and he called me very unjustly, a *poco curante* [caring, but only to a small degree – i.e. largely indifferent], and as I did not understand what he meant, it seemed to me a fearful reproach.[17]

In May 1823 Erasmus told Darwin, 'I have got a few Devils toe nails [belemnites – fossilized molluscs] for you…[18]' (i.e. which Darwin could add to his collection).

About a month later, Darwin's younger sister Emily, told him

> You have no idea how I long to seen you again my dear Charles … . How snug the Laboratory will be in Winter!! How does Mineralogy, Botany, Chemistry and Entomology [the study of insects] go on?[19]

Two years later, in March 1825, Erasmus asked Darwin to 'look in ye English Systema Vegetab [*Systema Vegetabilum*, published in 1783, by Swedish naturalist and physician, Carolus Linnaeus, 1707–78] & copy me out the specific description of Pinus sylvestris [Scots pine]'.[20]

What can be deduced about the young Darwin, from what is known of his life to date? That he was intensely interested in the natural world; had an enquiring mind; loved to experiment, and was a great collector of specimens. However, academically, he had failed to live up to his father Robert's expectations of him as he noted.

> As I was doing no good at school, my father wisely took me away at a rather earlier age than usual, and sent me to Edinburgh University … .[21]

Both his father Robert, and his grandfather Erasmus, had studied medicine at Edinburgh University, and it was envisaged that Darwin would follow in their footsteps and become a doctor. The year was 1825 and he was now aged sixteen.

Chapter 4

Edinburgh

Having left Shrewsbury School, Darwin was rebuked by his father who told him, 'You care for nothing but shooting, dogs, and rat-catching and you will be a disgrace to yourself and all your family.[1]' The question now was, would Darwin prove his father wrong?

At Edinburgh, Darwin joined the university's Plinian Society (founded in 1823 and named after the Roman scholar Gaius Plinius Secundus, author of *Historia Naturalis*, an encyclopaedia of knowledge). The society

> consisted of students, and met in an underground room in the University for the sake of reading papers on natural science [that branch of science that deals with the physical world, e.g., physics, chemistry, geology, and biology[2]] and discussing them. I used regularly to attend, and the meetings had a good effect on me in stimulating my zeal and giving me new congenial acquaintances.[3]

He also became a member of the Royal Medical Society, 'but as the subjects [discussed] were exclusively medical I did not much care about them[4]'.

During the holidays Darwin read; took walking tours with friends, and shot 'black-game' (grouse) at Woodhouse, Shropshire (home of William Mostyn Owen) with William's son Arthur, and also at Maer Hall, Maer, Staffordshire, the home of his uncle Josiah Wedgwood (II).[5] Darwin declared that he was 'attached to and greatly revered my Uncle Jos [Josiah] … . He was the very type of an upright man, with the clearest judgement'.[6] Soon Darwin would have reason to be especially grateful to his uncle Josiah.

A feature of the Darwin family was the close and loving relationship which existed between Darwin and his siblings, as illustrated by the interest which they took in one another's activities. For example, in December 1825, Darwin's younger sister Catherine, referring to her brother's visits to the theatre, declared, 'What capital luck you are in, just to fall in with all the good London performers at Edinburgh, [John] Liston, Miss [Catherine] Stephens, and [William Charles] Macready … .'[7]

Darwin informed his elder sister Caroline, in January 1826 that 'Dr Duncan is so very learned that his wisdom has left no room for his sense… .'[8] This was a reference to Dr Andrew Duncan, Edinburgh University's Professor of the Theory of Medicine, to whom Darwin was not prepared to pay homage merely on account of the former's position and status. Later that month, Darwin wrote to his elder sister Susan, to say

> The whole family have been so very good in writing to me so often that I do not know whom to begin to thank first, so to save trouble I return my humble thanks to you all, from my Father down to little Katty [(Emily) Catherine, his sister].[9]

Caroline, writing from Shrewsbury on 22 March 1826, exhorted Darwin to read the *Holy Bible*

> & not only because you think it wrong not to read it, but with the wish of learning there what is necessary to feel & do to go to heaven after you die. I am sure I gain more by praying over a few verses than by reading simply – many chapters – I suppose you do not feel prepared yet to take the sacrament [the service of Christian worship at which bread and wine are consecrated and shared].[10]

And then, on a lighter note

> it made me feel quite melancholy the other day looking at your old garden, & the flowers which you used to be so happy watching. I think the time when you & Catherine were little children & I was always with you or thinking about you was the happiest part of my life & I dare say always will be.[11]

Five days later, Susan informed Darwin that their father Robert, had misgivings about him 'picking & [choosing] what lectures you like to attend as you cannot have enough information to know what may be of use to you'.[12] This was perfectly understandable. After all, it was Robert who was funding Darwin's education. However, the truth was that Darwin had become disillusioned with the course, as he subsequently revealed.

> I derived *no* advantage from the Lectures at Edinburgh, for they were infinitely dull … . I was disgusted at anatomy & attended only 2 or 3 lectures & this has been ever since an irreparable loss to me.

Here is another illustration of Darwin's character. It was what interested him that mattered! However, on the positive side

> Dr Grant [Robert Edmund Grant, physician and biologist] was not a Professor, but worked at zoology out of pure love, & his society was a great encouragement. I used to amuse myself with examining marine animals, but I did so solely for amusement.[13]

On 8 April 1826, Darwin told Caroline, 'I have tried to follow your advice about the Bible, what part of the Bible do you like best? I like the Gospels'.[14] To which Caroline replied approvingly,

> I must say dear Charles how glad I am you have been studying the Bible – I agree with [you] in liking St John's the best of the Gospels. I am very fond of that short Epistle of St James, as well as St Johns – I often regret myself that when I was younger & fuller of pursuits & high spirits I was not more religious … .[15]

By the spring of 1827 it was clear that Darwin believed himself to be 'on the wrong track', as far as his studies were concerned. He wrote:

> My father perceived, or he heard from my sisters, that I did not like the thought of being a physician, so he proposed that I should become a clergyman. I asked for some time to consider, as from what little I had heard or thought on the subject I had scruples about declaring my belief in all the dogmas of the Church of England; though otherwise I liked the thought of being a country clergyman. Accordingly I read with great care *Pearson on The Creed* [*An Exposition of the Creed*, by John Pearson, Bishop of Chester, published in 1676] and a few other books on divinity; and as I did not then in the least doubt the strict and literal truth of every word in the Bible, I soon persuaded myself that our Creed must be fully accepted.[16]

In other words, at this relatively early stage in Darwin's life, he felt able to take the *Holy Bible* on trust and embrace it unreservedly. However, this would not always be the case.

Darwin acquiesced, once again, to his father's wishes, and it was decided that he should attend Cambridge University, where, at Christ's College, he would read for a Bachelor of Arts degree (BA) as a first step towards becoming an Anglican clergyman. This would, of necessity, be for a so called 'pass degree', rather than for the more prestigious 'honours degree'. (In order to obtain an 'honours' degree it was necessary to sit for the mathematical Tripos – and mathematics was not Darwin's strong point.[17] Alternatively, an honours degree could be obtained by sitting for the classical Tripos, but only those possessed of high honours in mathematics were eligible to do this.)

Here it should be noted that had Darwin remained a Unitarian, and not become an Anglican then, as a 'dissenter', admission to Cambridge University would not have been open to him.

However, in order to sit for a BA pass degree a knowledge of the classics was required, and as Darwin had forgotten 'almost everything which I had learnt' at school on the subject, he was obliged to 'brush up' his classics with the help of a private tutor in Shrewsbury. This meant that his entrance to Christ's College, Cambridge was delayed from October 1827 until January 1828.[18]

Chapter 5

Cambridge

Whilst at Cambridge, Darwin, true to form, found time to indulge in those extra-curricular activities which were of particular interest to him. For example, he attended

> Henslow's lectures on Botany and liked them much for their extreme clearness, and the admirable illustrations …
>
> Henslow used to take his pupils, including several of the older members of the University, [on] field excursions, on foot or in coaches, to distant places or in a barge down the river, and lectured on the rarer plants and animals which were observed. These excursions were delightful.[1]

This was a reference to the Reverend John Stevens Henslow, Cambridge University's professor of botany, about whom Darwin subsequently wrote

> a circumstance which influenced my whole career more than any other … was my friendship with Professor Henslow. Before coming up to Cambridge, I had heard of him from my brother as a man who knew every branch of science, and I was accordingly prepared to reverence him. He kept open house once every week when all undergraduates and some older members of the University, who were attached to science, used to meet in the evening. I soon got, through Fox, an invitation, and went there regularly.[2]

Darwin's second cousin William D. Fox, 'a clever and most pleasant man', was at that time a fellow undergraduate at Christ's College. Like Darwin, he too was preparing to become a clergyman. It was Fox who introduced Darwin to entomology.[3] Said Darwin

> When I went to Cambridge … I worked like a slave at collecting. Henslow's Society was a great charm & benefit to me, & I liked much his Lecture on Botany.[4]

On a cultural note, Darwin made frequent visits to the Fitzwilliam Museum to admire the paintings in its art gallery. And, said he

> I … acquired a strong taste for music, and used very often to time my walks so as to hear on week days the anthem in King's College Chapel. But no pursuit at Cambridge was followed with nearly so much eagerness or gave me so much pleasure as collecting beetles.[5]

Also, Darwin described his 'passion for shooting and for hunting, and … riding across country'.

> I got into a sporting set, including some dissipated low-minded young men. We often used to dine together in the evening … and we sometimes drank too much, with jolly singing and playing at cards afterwards.
>
> I know that I ought to feel ashamed of days and evenings thus spent, but as some of my friends were very pleasant and we were all in the highest spirits, I cannot help looking back to these times with much pleasure.[6]

Darwin was clearly enjoying his life at Cambridge. However, on 12 June 1828 he wrote to Fox to say, 'I am dying by inches, from not having any body to talk to about insects'.[7] That October, Darwin had a request to make of his cousin

> I want to know the name of a butterfly, which you have got [i.e. in your collection], its wings are *most wonderfully jagged*, & of a reddish colour, [and] after an immense chase with all the servants in the house I at last captured it.[8]

Darwin told Fanny, daughter of William Mostyn Owen and sister of his friend Arthur later in October,

> I have not rode [a horse] since I saw you, but have *nobody* to *ride with* so *no wonder*!! [Also] Not one game of Billiards have I had since I play'd with you. I can get nobody to play with & am a[fr]aid for *want* of *practice* [that I] shall forget all my *fine strokes* … .[9]

Darwin enjoyed horse-riding and billiards, activities he could share with his friends.

Darwin's benign acquiescence to his father's wishes in respect of his future career did not prevent him from pursuing his own interests, whenever the opportunity arose. For example, on 29 October 1828, he tells Fox that he has 'struck up a friendship' with

Edinburgh entomologist the Reverend Frederick William Hope, who has invited him to

> bring over all my insects to Netley [Netley Hall being the Hope family seat which was situated about 5 miles south of Shrewsbury]. I could write all day about him … he has given me a great many water [beetles].[10]

Darwin wrote to Fox in glowing terms on Christmas Eve, about 'Mr Dash', his pointer.[11]

> He rises in my opinion hourly, & I would not sell him for a £5 pound note. – it would have excited your envy & spleen to have seen him on the scent of a covey of Birds … .

Darwin indicates to Fox that Robert Darwin shares his interest in natural history

> You cannot imagine how pleased my Father was with the Death-Head's [species of moth]; to use his own words 'If he himself had thought for a week he could not [have] picked out a present so acceptable'.[12]

On 26 February 1829, Darwin tells Fox that, the previous week, in London, he has met with the Reverend Hope, when the latter

> did little else but talk about & look at insects: his collection is most magnificent & he himself is the most generous of Entomologists he has given me about 160 new species … .

He also tells Fox that, whilst in London, he visited the Royal Institution, the Linnaean Society (for the study and promotion of all aspects of the biological sciences), the Zoological Gardens, '& many other places where Naturalists are gregarious.'[13]

In another letter to Fox, dated 15 March, Darwin refers to 'some very gay parties' which he has attended, and also 'a night's debauch'[14] – an indication that Darwin was far from narrow-minded in his outlook on life, which was by no means all work and no play.

Darwin wrote to Fox again on 23 April to commiserate with him over the death of his sister Mary.

> I feel most sincerely & deeply for you & all your family: But, at the same time, as far as anyone can, by his own good [principles] & religion be supported under such a misfortune, you, I am assured, well know where to look for such support.

And he goes on to refer to 'the pure & holy a comfort as [i.e. which] the Bible affords …'.[15]

Darwin tells Fox on 3 July that his father Robert, 'has got two Martens [weasel-like mammals], I believe both species ready-mounted at Mr Shaw's … .'[16] [Henry Shaw, taxidermist of Shrewsbury].

In May 1830 Darwin writes aggrievedly to Fox, whom he had hoped to meet with, from Cambridge

> I am very sorry to find that all our plans are likely to vanish into air. It is most unfortunate you being obliged to go with your Father & Mother to Cheltenham, for the weather is so fine, the beetles so numerous, our zeal so ardent that the Science would have received a benefit never to have been forgotten. [And Darwin ends his letter] I have seen a good deal of [Professor] Henslow lately & the more I see of him the more I like him[.] I have some thoughts of reading divinity with him the summer after next.[17]

Darwin describes how

> during the latter half of my time at Cambridge [I] took long walks with him [Henslow] on most days … . His knowledge was great in botany, entomology, chemistry, mineralogy, and geology. His strongest taste was to draw conclusions from long-continued minute observations. His judgement was excellent, and his whole mind well-balanced; but I do not suppose that any one would say that he possessed much original genius.[18]

Far from being overawed by the mighty academics of Cambridge, Darwin was intent on making his own mark!

'During my last year at Cambridge,' he tells us:

> I read with care and profound interest Humboldt's *Personal Narrative*. This work, and Sir J. Herschel's *Introduction to the Study of Natural Philosophy*, stirred up in me a burning zeal to add even the most humble contribution to the noble structure of Natural Science.

Little did Darwin know what an understatement this would prove to be![19]

> (The full title of German geographer, naturalist, and explorer Alexander von Humboldt's book, which he wrote in conjunction with Aimé Bonpland, was *A Personal Narrative of Travels to the Equinoctial Regions of America, During*

the Years 1799–1804. It was published in four volumes between 1814 and 1829. In it, he made observations on the native American people of South America, and on Aztec Art, and gave detailed descriptions of the region's flora and fauna. The full title of astronomer, mathematician, chemist, and philosopher Sir John F. W. Herschel's book, published in 1831, was *A Preliminary Discourse on the Study of Natural Philosophy*. It comprised an analysis of the history and nature of science.)

As he prepares for his final examinations, Darwin writes ruefully to Fox on 5 November.

> I have not stuck [i.e. impaled with a pin, prior to mounting] an insect this term & scarcely opened a [collection] case … . but really I have not spirits or time to do any thing. Reading makes me quite desperate, the plague of getting up [preparing and revising] for my subjects is next thing to intolerable. [And he ends the letter] I am very glad to hear … that you have at last heard of a [curacy - i.e. the post of curate] where you may read all the commandments without endangering your throat.[20]

Darwin congratulated Fox on 23 January 1831 for having passed his degree examination and expressed the hope that Fox will manage to obtain a curacy at Epperstone near Nottingham – which he did. As for Darwin, he confessed that

> During the three years which I spent at Cambridge my time was wasted as far as the academic studies were concerned, as completely as at Edinburgh and at school. [Nevertheless] in my last year I worked with some earnestness for my final degree of BA.[21]

In the event, in the Bachelor of Arts degree, 'ordinary' class, which he was awarded on 25 April,[22] Darwin was placed a very creditable tenth out of the 178 who had succeeded in passing the examination.[23]

Because Darwin's initial entry to Cambridge University had been delayed, he was obliged to remain at that institution for another two terms, even though he had passed his final examination. It was during this period that Professor Henslow 'persuaded me to begin the study of geology'. Darwin finally went down from (i.e. left) Cambridge on 16 June.[24]

By the summer it was clear that Darwin was preoccupied, not with theology and the prospect of becoming a clergyman, but with natural history and natural science. For example, on 11 July, he wrote to Professor Henslow to inform him that he had now obtained a clinometer[25] (an instrument for measuring the angle or elevation of slopes).

> I suspect, the first expedition I take, clinometer & hammer in hand, will send me back very little wiser & a good deal more puzzled than when I started. As yet I have only indulged in hypotheses; but they are such powerful ones, that I suppose, if they were put into action but for one day, the world would come to an end.

How prophetic these words would prove to be, in view of what was to follow![26]

The Reverend Adam Sedgwick, was Woodwardian Professor of Geology at Cambridge University. Said Darwin, when Sedgwick proposed a visit to North Wales, commencing in early August 1831, in order

> to pursue his famous geological investigations amongst the older rocks … Henslow asked him [Sedgwick] to allow me to accompany him. Accordingly he came and slept at my father's house' [i.e. 'The Mount', Shrewsbury, en route from Cambridge to Wales].

A local labourer had previously informed Darwin that he (the labourer) had discovered in 'an old gravel-pit near Shrewsbury … a large worn tropical Volute shell …'. (Marine mollusc of the genus *Voluta*). But when Darwin passed this information on to Sedgwick, the latter declared that the shell

> must have been thrown away by some one into the pit … [for] if really embedded there it would be the greatest misfortune to geology, as it would overthrow all that we know about the superficial deposits of the Midland Counties.[27]

Whereupon Darwin was

> utterly astonished at Sedgwick not being delighted at so wonderful a fact as a tropical shell being found near the surface in the middle of England. Nothing before had ever made me thoroughly realize … that science consists in grouping facts so that general laws or conclusions may be drawn from them.

Darwin also realized that because of his preconceived ideas, Sedgwick's mind was closed to the possible significance and importance of this finding.

In North Wales, Sedgwick and Darwin visited Cwm Idwal, a valley in Snowdonia. Here, whilst hunting for fossils, 'neither of us saw a trace of the wonderful glacial phenomena all around us; we did not notice the plainly scored rocks, the perched boulders, the lateral and terminal moraines'. This, said Darwin, was a 'striking instance [of] how easy it is to overlook phenomena, however conspicuous …[28]'.

* * *

For his degree Darwin had been examined in the works of Homer (Greek poet and man of letters), Virgil (Roman poet), Euclid (Greek mathematician), and in arithmetic and algebra. However, and not surprisingly, the syllabus was heavily weighted towards theology, viz the required reading included philosopher and physician John Locke's *An Essay Concerning Human Understanding*, and theologian William Paley's *View of the Evidences of Christianity* (1794).[29]

Chapter 6

John Locke and William Paley

Anyone who believed that the syllabus for Darwin's BA degree would entail a comparative and objective study of various world religions, or even a genuine debate on the merits or demerits of Christianity itself, would be sadly disappointed. The syllabus was specifically designed to include only Anglican theology, as put forward by theologists/ philosophers whose minds were already made up.

John Locke (1632–1704)

Chapter X of Book IV of John Locke's *An Essay Concerning Human Understanding* (1690) is entitled 'Of Our Knowledge of the Existence of a God'. It reads as follows

> 1. *We are capable of knowing certainly that there is a God.* Though God has given us no innate ideas of himself, though he has stamped no original characters on our minds, wherein we may read his being; yet having furnished us with those faculties our minds are endowed with, he hath not left himself without witness: since we have sense, perception, and reason, and cannot want a clear proof of him, as long as we carry ourselves about us.

In other words, the very fact of man's existence is proof of the existence of God. Locke then takes his argument a step further

> 2. *For man knows that he himself exists.* I think it is beyond question, that man has a clear idea of his own being; he knows certainly he exists, and that he is something.
>
> This, then, I think I may take for a truth, which every one's certain knowledge assures him of, beyond the liberty of doubting, viz that he is something that actually exists.
>
> 3. He knows also that nothing cannot produce a being; therefore something must have existed from eternity.

In other words, who else but God could have created man?

> 4. *And that eternal Being must be most powerful.*
>
> 5. … a man finds in himself perception and knowledge. We have then got one step further; and we are certain now that there is not only some being, but some knowing, intelligent being in the world. [For it was impossible] that things wholly void of knowledge, and operating blindly, and without any perception, should produce a knowing being … .

Therefore, that is, as man is intelligent, God must also be intelligent.

> 6. Thus, from the consideration of ourselves, and what we infallibly find in our own constitutions, our reason leads us to the knowledge of this certain and evident truth, That there is an eternal, most powerful, and most knowing Being; which whether any one will please to call God, it matters not. [And finally] From what has been said, it is plain to me we have a more certain knowledge of the existence of a God, than of anything our senses have not immediately discovered to us. Nay, I presume I may say, that we more certainly know that there is a God, than that there is anything else without us.[1]

Despite his assuredness, Locke produces no evidence to support his theory of the existence and nature of God, his arguments being based solely on supposition.

William Paley (1743–1805)

In the preface to *A View of the Evidences of Christianity*, published in 1746, Paley declared that he was moved to write the book because of his 'earnest wish to promote the religious part of an academic education'.[2] Said he

> I deem it unnecessary to prove that mankind stood in need of a Revelation [the disclosure of knowledge by a divine], because I have met with no serious person who thinks that even under the Christian revelation we have too much light, or any degree of assurance which is superfluous. I desire moreover that in judging of Christianity it may be remembered, that the question lies between this religion and none: for, if the Christian religion be not credible, no one, with whom we have to do, will support the pretensions of any other.[3]

In other words, man requires a 'god', the Christian god provides him with both enlightenment and certainty, and all other religions are irrelevant. Paley then asks, 'In what way can a revelation be made but by miracles?' To this he answers, 'In none which we are able to conceive'.[4]

> We are in possession of letters written by St Paul himself upon the subject of his ministry, and either written during the period which the history comprises, or, if written afterwards, reciting and referring to the transactions of that period.[5]

This was a reference to the Christian missionary Paul, otherwise known as Saul of Tarsus (in modern south-east Turkey). His dates are believed to be circa AD1–10 to circa AD64–68, and there is no evidence that he ever met Jesus Christ, nor did he ever claim to have done so.

In respect of Christ's 'associates', Paley declared

> From the clear and acknowledged parts of the case, I think it to be likewise in the highest degree probable, that the story, for which these persons voluntarily exposed themselves to the fatigues and hardships which they endured, was a *miraculous* story ...[6]

Here, Paley appears to be arguing that because the authors of the gospels, Matthew, Mark, Luke, and John, had taken risks and endured hardships, this made it highly likely that their accounts were true. In respect of the miracles themselves, he declares

> We have this detail from the fountainhead, from the persons themselves; in accounts written by eye-witnesses at the scene, by contemporaries and companions of those who were so; not in one book but four [i.e. the gospels] each containing enough for the verification of the religion, all agreeing in the fundamental parts of the history.[7]

'In viewing the detail of miracles recorded in these books, we find every supposition negative, by which they can be resolved into fraud or delusion.' Each account was 'contemporary, it was published upon the spot'.[8]

As for the gospels themselves, said Paley

> No stronger evidence of the truth of a history can arise from the situation of the historian than what is here offered. The authors of all the histories lived at the time and upon the spot.[9]

In other words, the accounts provided by the gospel writers were based on first-hand experience, and, in Paley's words, the fact that the 'different narratives' of the gospels varied from one another, was evidence, such 'as to repel all suspicion of confederacy [i.e. collusion between these writers] ...'.[10]

The truth is that despite centuries of research by Christian scholars, it has proved impossible to date the lives of the gospel writers. Furthermore, best estimates give the dates for the gospels as follows: St Matthew, sometime between AD50–75; St Mark, AD late 30s–75; St Luke, AD60–90, and St John, AD50–85. As Christ is alleged to have died circa AD30–36, it appears that despite Paley's assertion, none of the gospels were written during Christ's lifetime. (And, as far as is known, none of the Gospel writers ever claimed that this was the case.)

Paley proceeds to quote from a host of 'learned scholars' from all over the western world, who lived in the succeeding centuries (notably the fourth), and who have affirmed the authenticity of the Christian gospels.[11] For example, he quotes Chrysostom 'who lived near the year 400' and who declared, 'The general reception of the Gospels is a proof that their history is true and consistent'.[12]

However, Paley does admit that some doubted not only the authority of the gospels, but also their authorship; whilst others accused the early Christians of altering the text. Such people, who include the Valentinians and the Carpocratians, are dismissed by Paley as being 'heretics'.[13]

The gospel writers and other contemporary witnesses, said Paley

> could not be deceivers. By not only bearing testimony, they might have avoided all their sufferings, and have lived quietly. Would men in such circumstances pretend to have seen what they never saw; assert facts which they had no knowledge of; go about lying …[14]

Paley's stance is only what was to be expected from one who held the post of Archdeacon of Carlisle. The question is, despite the rhetoric – defined as the art of effective or persuasive speaking or writing[15] - of Locke and Paley, would Darwin continue to accept Christianity as being an authentic, evidence-based religion?

Chapter 7

A Proposition

Before any thoughts of becoming a clergyman could be entertained, Darwin received, on 21 August 1831, a letter from Henslow which would change his life. It read as follows:

> I shall hope to see you shortly fully expecting that you will eagerly catch at the offer which is likely to be made you of a trip to Terra del Fuego & home by the East Indies – I have been asked by Peacock [George Peacock, Fellow of Trinity College, Cambridge and lecturer in mathematics] ... to recommend [to] him a naturalist as companion to Capt [Robert] FitzRoy employed by the Government to survey the S. extremity of America – I have stated that I consider you to be the best qualified person I know of who is likely to undertake such a situation – I state this not on the supposition of yr being a *finished* Naturalist, but as amply qualified for collecting, observing, & noting any thing worthy to be noted in Natural History. The voyage is to last 2 yrs ...[1]

Two days later Peacock himself wrote to Darwin to inform him that

> The expedition is entirely for scientific purposes & the ship will generally wait your leisure for researches in natural history &c ... . The Admiralty are not disposed to give a salary, though they will furnish you with an official appointment ... .

In the event Darwin's 'appointment' was not made official, so having accepted Peacock's offer, he gave himself the unofficial title 'Naturalist to the *Beagle*' (a reference to HMS *Beagle*, the vessel in question), or alternatively, 'Naturalist to the Expedition'.[2]

Darwin's father Robert, however, was initially opposed to the idea, as the following letter from him to his (Robert's) brother-in-law Josiah Wedgwood (II), indicates. He begins the letter by expressing satisfaction that Wedgwood, who was evidently a patient of Robert's medical practice, has benefited from the 'turpentine pills' which he has recently prescribed for the former's 'bowel' [i.e. intestinal] disorder. He then proceeds to discuss with Wedgwood his son Charles's proposed 'voyage of discovery'.

> I strongly object to it [on var]ious grounds, but I will not detail my reasons [in order] that he may have your unbiased opinion on the subject, & if you think differently from me I shall wish him to follow your advice.[3]

On 31 August 1831 Darwin wrote to his father saying,

> I am afraid that I am going to make you again very uncomfortable. But upon consideration, I think you will excuse me once again stating my opinions on the offer of the Voyage. My excuse & reason is… the different way all the Wedgwoods view the subject from what you & my sisters do.

However, if the answer was still 'No', then, said Darwin

> I should be most ungrateful if I did not implicitly yield to your better judgement & to the kindest indulgence which you have shown me all through my life.[4]

On that same day Josiah Wedgwood himself wrote to Robert Darwin to say

> as you have desired Charles to consult me I cannot refuse to give the result of such consideration as I have been able to give it. Charles has put down what he conceives to be your principal objections & I think the best course I can take will be to state what occurs to me upon each of them.

There were, in fact, eight objections (which Darwin had previously laid before Josiah Wedgwood): the most important of which were that the expedition would be

> (1) Disreputable to my [son Charles's] character as a Clergyman hereafter.
> (8) That it would be a useless undertaking.

To these points, Wedgwood's response was as follows

> 1 – I should not think that it would be in any degree disreputable to his character as a clergyman. I should on the contrary think the offer honourable to him, and the pursuit of Natural History, though certainly not professional [i.e. not one of the three learned professions – theology, law or medicine], is very suitable to a Clergyman.

> 8 – The undertaking would be useless as regards his [intended] profession [of clergyman, Wedgwood conceded] but looking upon him as a man of enlarged curiosity, it affords him such an opportunity of seeing men and things as happens to few.[5]

Finally, on 1 September 1831, Robert Darwin relented, and in a letter to Josiah Wedgwood, declared

> Charles has stated my objections quite fairly & fully [i.e. to Wedgwood] – if he still continues in the same mind after further enquiry, I will give him all the assistance in my power. Many thanks for your kindness – yours [affectionately] R W Darwin

On that same day, Darwin wrote to Francis Beaufort, Hydrographer of the Navy, to say that his father Robert, had 'given his consent & therefore if the appointment is not already filled up, I should be very happy to have the honour of accepting it'.[6]

From London, Darwin wrote to his sister Susan four days later to say that he had met Captain FitzRoy and

> it is no use attempting to praise him as much as I feel inclined to do, for you would not believe me. One thing I am certain of nothing could be more open & kind than he was to me.[7]

HMS *Beagle* was a Cherokee-class, brig-sloop built for the Royal Navy and launched on 11 May 1820. However, she was to spend her working life employed at Woolwich Dockyard as a survey barque. The first expedition (May 1826 to October 1830) upon which *Beagle* embarked was to Patagonia and Tierra del Fuego, in order to make a hydrographic survey of the region. In the latter stages of this expedition Flag Lieutenant Robert FitzRoy was made temporary captain of the ship. In her second expedition (December 1831 to October 1836) – the aim of which was to complete the survey of the South American coast – FitzRoy was appointed *Beagle*'s Commander and Surveyor.

In that same letter to Susan, Darwin describes HMS *Beagle* as having a complement of '60 men [and] 5 or 6 officers &c'.[8] (The ship was, in fact, ninety feet in length and 242 tons burthen – i.e. tonnage.)

On 6 September he wrote to Susan again, telling her to request Edward (unidentified, but presumably a member of his father's household staff) 'to send me up in my carpet bag, my slippers, a pair of lightish walking shoes. My Spanish books [presumably relating to Spanish influence in South American]: my new microscope … [and] my geological compass', together with a volume on the subject of taxidermy.[9]

Darwin tells Susan three days later, that the previous day, 8 September, he had been to see the procession for the coronation of King William IV (who had acceded to the throne in the previous year, and would reign until 1837).[10]

On the 14th Darwin informed Susan that:

> The vessel is a very small one; three-masted; & carrying 10 guns: but every body says it is the best sort for our work, & of its class it is an excellent vessel: new, but well tried, & ½ again [i.e. 50 per cent greater than] the usual strength. [However] The want of room is very bad but we must make the best of it.[11]

Darwin told Caroline on 12 November

> Everything here is most prosperous; the Beagle now looks something like a ship They have just painted her and in a weeks time the men will live on board. No Vessel has ever been fitted at all on so expensive a scale from Plymouth ...[12]

In his autobiography, Darwin gave further details of how Josiah Wedgwood had persuaded his father Robert, to change his mind about the *Beagle* voyage. The long-suffering Robert, showing considerable forbearance towards his son, had told him, 'If you can find any man of common-sense who advises you to go I would give my consent.' Darwin also records that 'My father always maintained' that Josiah Wedgwood 'was one of the most sensible men in the world ....' What more natural, therefore, that Darwin should seek his uncle's advice on the subject, knowing that if Josiah approved, the battle was won? Darwin senior had been 'hoist with his own petard'!

Chapter 8

The Voyage of HMS *Beagle*

HMS *Beagle* finally sailed from Plymouth on 27 December 1831. During her voyage Darwin would see life on Earth in all its astonishing diversity and participate in what was, arguably, to be the greatest and most fruitful scientific voyage to date in the history of the world. Said he

> The object of the expedition was to complete the survey of Patagonia [a region stretching northwards from Tierra del Fuego for 1,000 miles, and since 1881 divided between Argentina and Chile] and Tierra del Fuego [an archipelago off the southernmost tip of South America] – commenced under Captain [Philip Parker] King in 1826 and ending in 1830, to survey the shores of Chile, Peru, and some islands in the Pacific, and to carry a chain [sequence] of chronometrical measurements round the world.[1]

In the poop cabin, where Darwin worked and slept in his hammock, was a library which contained in excess of a hundred volumes.[2] For him, favourite authors included French palaeontologist Alcide d'Orbigny, and German naturalists and travellers Alexander von Humboldt and Christian Ehrenberg. Having thus learned about the natural history of various countries of the world – even though he had not visited them – this would enable him to compare their flora, fauna, geology, etc., with that of South America.

> My education in fact began on board the *Beagle*. I remember nothing previously which deserved to be called education except some experimental work at chemistry when a school-boy with my Brother [Erasmus Alvey Darwin].[3]

On his previous *Beagle* expedition (1826 to 1830), Captain FitzRoy had brought four indigenous natives from Patagonia back to England, with the object of educating them at his own expense. One, nicknamed 'Boat Memory', had died of smallpox soon after their arrival. However the other three: 'York Minster', James ('Jemmy') Button, and 'Fuegia Basket' (a female), were currently aboard *Beagle* and awaiting repatriation.[4]

An account of the voyage is to be found in Darwin's *Journal and Remarks* – otherwise known as *The Voyage of the Beagle* – published in 1839.

> 16 January 1832. … we anchored at Porto Praya, in St Jago, the chief island of the Cape de Verd archipelago [Atlantic islands of Cape Verde] … the geology of this island is the most interesting part of its natural history.[5]

On 10 February, Darwin wrote to his father, Robert, to say

> What may appear quite paradoxical to you, is that I *literally* find a ship (when I am not sick) nearly as comfortable as a house. It is an excellent place for working & reading, & already I look forward to going to sea, as a place of rest, in short my home.[6]

However, he had spoken too soon, for in February/March, from Salvador (Salvador de Bahia), Brazil, he wrote to his father again, to confess that 'the misery I endured from sea-sickness' as the ship passed through the Bay of Biscay was 'far far beyond what I ever guessed at'.[7]

Captain FitzRoy was a person of considerable ability, who, following the present expedition, would be awarded the Royal Geographical Society's gold medal. In later years, he would serve as a Member of Parliament and Meteorological Statist to the Board of Trade. In the latter capacity, and as the inventor of various types of barometer, he insisted that such instruments were installed at all ports, in order a), that they could be consulted by captains prior to setting out to sea and b), that such data could be relayed back to himself, for analysis. It was FitzRoy who coined the phrase 'forecasting the weather' – hence the term 'weather forecast'.

FitzRoy was also to write a *Narrative* of the present voyage, in which he declared

> Undoubtedly the worst wind, next to a hurricane, which a vessel can encounter, is a violent 'white squall,' so called because it is accompanied by no cloud or peculiar appearance in the sky, and because of its tearing up the surface of the sea, and sweeping it along so as to make a wide sheet of foam.

Such squalls gave no warning of their approach

> but by consulting a good barometer or sympiesometer [lightweight barometer], and frequently watching the surface of the sea itself, even a white squall may

> be guarded against in sufficient time.[8]

Meanwhile, Darwin's relationship with the captain of HMS *Beagle* was not always a happy one.

> We had several quarrels; for instance, early in the voyage at Bahia, in Brazil, he defended and praised slavery, which I abominated …[9]

Darwin had shown himself to be a person of unswerving principle, who was not afraid to speak his mind even if this meant, in this case, offending his captain, who had it in his power to make life exceeding difficult for the young naturalist, had he chosen to do so.

HMS *Beagle* arrived at San Salvador in north-eastern Brazil on 29 February. Here, Darwin declared

> The day has passed delightfully. Delight itself, however, is a weak term to express the feelings of a naturalist who, for the first time has wandered by himself in a Brazilian forest.[10]

During the period 4 April to 5 July, *Beagle* was engaged in an oceanic survey of the Brazilian coast in the vicinity of Rio de Janeiro. However, in his narrative of the voyage,[11] Captain FitzRoy indicates that the mission was not without its tragedy.

> On this passage one of our seamen died of a fever, contracted when absent from the *Beagle* with several of her officers, on an excursion to the interior part of the extensive harbour of Rio de Janeiro. One of the ship's boys, who was in the same party, lay dangerously ill, and young [Charles] Musters [Volunteer 1st Class] seemed destined to be a victim to the deadly fever.
>
> I found that the [River] Macacú was notorious among the natives as being often the site of pestilential malaria, fatal even to themselves. How the rest of our party escaped, I know not; for they were eleven or twelve in number, and occupied a day and night in [i.e. on] the river.
>
> As far as I am aware, the risks, in cases such as these, is chiefly encountered by sleeping on shore, exposed to the air on or near the low banks of rivers, in woody or marshy places subject to great solar heat. Those who sleep in boats, or under tents, suffer less than persons sleeping on shore and exposed … .[12]

On 19 May, said FitzRoy, Musters 'ended his short career'. (FitzRoy also stated that 'The boy Jones … died the day after our arrival in Bahia.'[13] This was Bahia Blanca, Argentina, which *Beagle* arrived at on 4 August.)

Darwin gave a similar account of the tragedy. When the *Beagle* sailed from Rio, he said, three of the party who had explored the River Macacú 'were ill with fever', including Seaman Morgan, 'Boy Jones' and Charles Musters.

> Macacú has been latterly especially notorious for fevers: how mysterious and how terrible is their power. It is remarkable that in almost *every* case, the fever appears to come on several days after returning into the pure atmosphere.[14]

Meanwhile, on 8 April, Darwin, in company with an Englishman with whom he had become acquainted, embarked on a 100-mile overland journey on horseback into the interior, where he made measurements of temperature and rainfall, studied fireflies, fungi, and tree ferns, and collected insects in the forest.

In May, Darwin admits to being bewildered by the host of opportunities which the *Beagle* voyage is affording him, when he writes from Botafogo Bay, Rio de Janeiro to his cousin William Fox, to say

> My collections go on admirably in almost every branch. As for insects, I trust I shall send an host of [hitherto] undescribed species to England, [However, at present] I am entirely occupied with land animals, as the beach is only sand … . But Geology carries the day; it is like the pleasure of gambling, speculating on first arriving [as to] what the rocks may be … .[15]'

In May/June, Darwin wrote to Henslow to say

> I have so many things to write about, that my head is as full of oddly assorted ideas, as a [specimen] bottle on the table is with animals.[16] [Darwin preserved his organic specimens –such as fungi – in spirit, or failing that, in gin and brine.]
>
> One great source of perplexity to me is an utter ignorance whether I note the right facts & whether they are of sufficient importance to interest others.
>
> Tell Prof. Sedgwick he does not know how much I am indebted to him for the Welch [Welsh] expedition. – it has given me an interest in geology, which I would not give up for any consideration.

And he promises that when *Beagle* arrives at Montevideo – capital and chief port of Uruguay – he will send Henslow 'a box' (i.e. a packing case) of specimens.[17]

Writing from Rio de Janeiro, on 5 July, Darwin tells his sister, Catherine:

> The geography of this country is as little known as [the] interior of Africa. We have been 3 months here: & most undoubtedly I well know the glories of a Brazilian forest. The number of undescribed animals I have taken [collected] is very great … . I attempt class after class of animals, so that before very long I shall have a notion of all.[18]

In other words, rather than concentrating on a limited number of species, Darwin's approach was to give blanket coverage to as much of the animal kingdom as was humanly possible.

> 26 July 1832. We anchored at Montevideo [capital of Uruguay, and] stayed ten weeks at [the town of] Maldonado ... situated on the northern bank of the [Rio de la] Plata [River Plate].

It was at Montevideo that Captain FitzRoy met English-born landscape painter Conrad Martens, who was promptly engaged to replace *Beagle*'s ship's artist Augustus Earle, who had fallen ill. Darwin and Martens struck up an enduring friendship.

During this time 'a nearly perfect collection of the animals, birds, and reptiles was procured'. From here Darwin undertook a seventy-mile overland excursion to the River Polanco, pausing to study, en route, the dress and habits of the Gauchos (cowboys of the South American pampas).[19]

In July/August, Darwin writes to Henslow to say that 'to day I have been out & returned like Noah's ark – with all sorts'.[20]

Darwin's letter to Henslow of 26 October showed that his prolonged absence from home was beginning to take its toll.

> On board the Ship, everything goes on as well as possible, the only drawback is the fearful length of time between this [day] & [the] day of our return. I do not see any limits to it: one year is nearly completed & the second will be so before we even leave the East coast of S. America. And then our voyage may be said really to have commenced. I know not, how I shall be able to endure it. The frequency with which I think of all the happy hours I have spent at Shrewsbury & Cambridge, is rather ominous.[21]

In October/November, Darwin tells his sister Caroline

> It is now nearly four months, since I have received a letter ... . I am glad the journal arrived safe ... . I feel it is of such consequence to my preserving a just recollection of the different places we visit.

This was a reference to the journal in which Darwin was keeping an account of the voyage.[22]

To be a member of such a close family, and to be separated from that family by seemingly endless time and space, and to know that the voyage would continue, seemingly indefinitely, must, to Darwin in his more reflective moments, have made him extremely homesick.

> 17 December 1832. In the afternoon we anchored [at Tierra del Fuego] in the Bay of Good Success.[23]
> 21 December 1832. The *Beagle* … [at] about three o'clock doubled [sailed around] the weather-beaten Cape Horn.[24]

In January 1833, Darwin received a letter from Henslow at Cambridge to say, 'So far from being disappointed with the Box [of specimens] – I think you have done wonders … . Most of the plants are very desirable to *me*'.[25] The ship now near doubled back and returned to the Atlantic Ocean.

> 1 March 1833. … the *Beagle* anchored in Berkeley Sound in East Falkland Island.[26]

During this and the following month, Darwin appraised Caroline of the terrible sea conditions encountered off Cape Horn.

> We doubled Cape Horn on a beautiful afternoon; it was however the last we were doomed [destined] to have for some time. After trying to make head [way] against the Westerly gales we put into a cove near the Cape. Here we experienced some tremendous weather; the gusts of wind fairly tear up the water & carry clouds of spray. We again put to sea, with no better success, gales succeed gales, with such short intervals, that a ship can do nothing.[27]

In April Darwin informed Henslow that:

> The geology of this part of Tierra del [Fuego] was, as indeed every place is, to me very interesting. [However] In Zoology during the whole [cruise], I have done little; the Southern ocean is nearly as sterile as the continent it washes.[28]

In May/July, Darwin informed Catherine that although he was tempted to 'make a bolt across the Atlantic to good old Shropshire', he had decided against it.

> I have worked very hard (at least for me) at Nat History & have collected many animals & observed many geological phenomenon: & I think it would be a pity having gone so far, not to go on & do all in my power in this my favourite pursuit; & which I am sure, will remain so for the rest of my life.[29]

Darwin had made the correct decision, for had he but known it, a treat such as any student of natural history would have relished beyond measure, was just around the corner.

By 18 July, when he wrote to Henslow from the Rio de la Plata, to which the *Beagle* had returned, Darwin's spirits had lifted.

> I am ready to bound for joy at the thoughts of leaving this stupid, unpicturesque side of America. When Tierra del F. is over, it will be all Holidays. And then the very thoughts of the fine [corals], the warm glowing weather, the blue sky of the Tropics is enough to make one wild with delight.[30]

> 24 July 1833. The *Beagle* sailed from Maldonando, and on August the 3rd she arrived at the mouth of the Rio Negro [Argentina].

Here, Darwin took note of the 'salinas', each of which 'during the winter ... consists of a shallow lake of brine, which in summer is converted into a field of snow-white salt. This, the Patagones [inhabitants of Patagonia] harvest and use for preserving cheese and meat'.[31]

From the Rio Negro, Darwin embarked on another overland journey, this time one of several hundred miles. Said he

> I hired a gaucho to accompany me on my ride to Buenos Aires ... [capital city of Argentina]. The distance ... is about four hundred miles, and nearly the whole way through uninhabited country.[32]

For the journey Darwin was placed under the protection of General Juan Manuel de Rosas, Governor of the Province of Buenos Aires and his army, which was charged with exterminating the 'wandering tribes of horse Indians, which have always occupied the greater part of this country, having of late much harassed the outlying *estancias* [rural estates, or ranches]'.[33] For such a journey, considerable stamina was required, even though it was undertaken on horseback. Sadly, stamina was a quality which Darwin would not always possess.

The village of Bahia Blanca (Argentina), said Darwin, consists of

> a few houses and the barracks for the troops [which were] enclosed by a deep ditch and fortified wall.[34] The *Beagle* arrived here on the 24th August, and a week afterwards sailed for the [Rio de la] Plata [or 'River Plate', on which Buenos Aires is situated]. With Captain Fitz Roy's consent I was left behind to travel by land to Buenos Aires.[35]

Darwin now proceeds to give a description of the geology of the region, and in particular of 'the number and extraordinary character of the remains of gigantic land animals'.[36] They included the *megatherium*, *megalonyx*, *scelidotherium*, *mylodon*, *pachydermatous*, and *toxodon*.[37] 'The remains of these nine great quadrupeds, and many detached [fossilized] bones were found embedded on the beach within the space of about 200 yards square.'[38] As for living animals, ostriches and armadillos provided interesting subjects for study.

As regards the constant battles between General Rosas's troops and the 'wild Indians', Darwin declared, 'Everyone here is fully convinced that this is the most just war, because it is against barbarians.' Darwin, however, considered both sides to be equally barbaric: 'Who would believe in this age that such atrocities could be committed in a civilized Christian country?[39] The warfare is too bloody to last; the Christians killing everything Indian, and the Indians doing the same to the Christians.'[40]

> 20 September 1833. We arrived by the middle of the day at Buenos Aires.[41]

Here, another marathon journey awaited him, for seven days later he wrote

> In the evening I set out on an excursion to St [Santa] Fe, which is situated nearly three hundred English miles from Buenos Aires, on the banks of the [river] Parana.[42]

By comparing the various animals which inhabited the continents of North and South America, both currently and prehistorically (i.e. from their bony remains), Darwin was able to deduce that the two continents were once

> much more closely related in the character of their terrestrial inhabitants than they now are … I know of no other instance where we can almost mark the period and manner of the splitting up of one great region into two well-characterized zoological provinces. [Furthermore] The geologist, who is fully impressed with the vast oscillations of level which have affected the Earth's crust within late periods, will not fear to speculate on the recent elevation of the Mexican Platform, or, more probably, on the recent submergence of land in the West Indian archipelago, as the cause of the present zoological separation of North and South America.[43]

On 12 October, Darwin declared, 'I had intended to push my excursion further, but not being quite well, I was compelled to return by a *balandra*, or one-masted vessel of about a hundred tons burden, which was bound to Buenos Aires.'[44]

From Buenos Aries Darwin took a packet (passenger boat) for Montevideo, capital and chief port of Uruguay. From here, he embarked on yet another journey into the interior; this one being of three weeks duration.

> 6 December 1833. The *Beagle* sailed from the Rio Plata, never again to enter its muddy stream.[45]

One evening, when HMS *Beagle* was about ten miles offshore, Darwin observed 'vast numbers of butterflies.[46] Our course was directed to Port Desire [Argentina], on the coast of Patagonia', at which place they arrived on 23 December. Here, Darwin observed that 'the guanaco, or wild llama, is the characteristic quadruped of the plains of Patagonia; it is the South American representative of the camel of the East'.[47]

> 9 January 1834. Before it was dark the *Beagle* anchored in the fine, spacious harbour of Port St Julian, situated about one hundred and ten miles to the south of Port Desire.[48] [Here] 'a good-sized fly (Tabanus) was extremely numerous, and tormented us by its painful bite. The common horsefly, which is so troublesome in the shady lanes of England, belongs to this same genus.'[49]

(Horseflies of the genus *Tabanus* are known to be potential vectors of anthrax, worms, and trypanosomes – a trypanosome being a single-celled parasitic protozoan with a trailing flagellum, and a flagellum being a microscopic whip-like appendage which enables it to swim.[50] The significance of this, for Darwin, will be discussed later.)

With the fossilized bones which he had discovered in mind, Darwin declared

> It is impossible to reflect on the changed state of the American continent without the deepest astonishment. Formerly it must have swarmed with great monsters: now we find mere pigmies, compared with the antecedent, allied races. Certainly, no fact in the long history of the world is so startling as the wide and repeated extermination of its inhabitants.
>
> We do not steadily bear in mind how profoundly ignorant we are of the conditions of existence of every animal, nor will we remember that some check is constantly preventing the too rapid increase of every organized being left in a state of nature. The supply of food, on an average, remains constant, yet the tendency in every animal to increase by propagation is geometrical … .[51]

In this latter comment, Darwin was voicing the warnings of economist and clergyman Thomas Robert Malthus (1766–1834), who will be discussed later. Yet, in his own mind, the seeds of the great theory for which he was to become famous were beginning to germinate.

In February, when *Beagle* was anchored in the Bay of Port Famine in the Strait of Magellan (navigable sea route between mainland South America and Tierra del Fuego), Darwin undertook the ascent of the 2,600-foot-high Mount Tarn.[52] However, as far as flora and fauna were concerned, he was disappointed.

> The zoology of Tierra del Fuego, as might have been expected from the nature of its climate and vegetation, is very poor.[53] The absence of any species

> whatever in the whole class of reptiles is a marked feature in the zoology of this country, as well as in that of the Falkland Islands.[54]
> 28 February 1834. … the *Beagle* anchored in a beautiful little cove at the eastern entrance of the Beagle Channel [strait separating islands in the Tierra del Fuego archipelago, and named after HMS *Beagle* following her first voyage to south America].[55]

16 March found HMS *Beagle* again at the Falkland Islands. In that month John M. Herbert, a friend from his Cambridge days, wrote to Darwin to say

> At the present moment everybody is talking about & the London Papers are full of, a petition from all the good men & true at Cambridge in [favour] of Dissenters being admitted to the University.[56]

In the Strait of Magellan Darwin was very impressed with the Patagonians who, with 'their large Guanaco mantles [llama-skin cloaks] & long flowing hair, have a very imposing appearance'. They were 'semi-civilized', and many of them could speak 'a little Spanish'.[57]

> 13 April 1834. 'The *Beagle* anchored within the mouth of the Santa Cruz [river, south-east Argentina]', and her crew proceeded to explore its course using three whale boats [a 'whale boat' being a ship's lifeboat and utility boat].

Said Darwin, 'Patagonia, poor as she is in some respects, can however boast of a greater stock of small rodents than perhaps any other country in the world'.[58]

On 18 April Captain FitzRoy declared

> Much of my own uneasiness was caused by reading works written by geologists who contradict, by implication, if not in plain terms, the authenticity of the Scriptures … . But [for] men who, like myself formerly, are willingly ignorant of the Bible, and doubt its divine inspiration, I can only have one feeling – sincere sorrow.[59]

In his *Narrative*, which contains numerous references to the *Bible*, FitzRoy reveals that he is looking for God's handiwork in all that he sees before him. Also, that he is a Creationist who entertained absolutely no doubts as to the validity of Biblical teaching. Says he:

> Have we a shadow of ground for thinking that wild animals or plants have improved since their creation? Can any reasonable man believe that the first of

> a race, species, or kind, was the most inferior? Then how for a moment could false philosophers, and those who have been led away by their writings, imagine that there were separate beginnings of savage races, at different times, and in different places? Yet I may answer this question myself; for until I had thought much on the subject, and had seen nearly every variety of the human race, I had no reason to give in opposition to doubts excited by such sceptical works, except a conviction that the Bible was true, that in all ages men had erred, and that sooner or later the truth of every statement contained in that record would be proved.[60]

Here is an indication of the difference in approach between Fitzroy and Darwin, who was three years his junior. The captain, who was also interested in the natural world, asked not how naturally occurring phenomena could be explained, but how they could be explained within the context of Biblical teachings. Darwin, on the other hand, was inhibited by no such constraints.

> 1 June 1834. We anchored [again] in the fine bay of Port Famine.
> 10 June 1834. In the morning we made the best of our way into the open Pacific.[61]

As *Beagle* voyaged northward along the west coast of Chile, Darwin observed that

> almost every arm of the sea, which penetrates to the interior higher chain [of mountains], not only in Tierra del Fuego, but on the coast for 650 miles northwards, is terminated by 'tremendous and astonishing glaciers'.[62]

In June/July, Darwin told Catherine, 'I collect every living creature, which I have time to catch & preserve; also some plants'.[63]

> 23 July 1834. The *Beagle* anchored late at night in the Bay of Valparaiso, the chief sea port of Chile.[64]

Here, ship's artist Martens left the ship and sailed for Australia. On that same day Darwin wrote to Charles T. Whitley, another friend from his Cambridge days (and John M. Herbert's cousin) to say

> I have seen nothing, which more completely astonished me, than the first sight of a Savage; it was a native Fuegian [inhabitant of Tierra del Fuego] his long hair blowing about, his face besmeared with paint. There is in their countenances, an expression, which I believe to those who have not seen it,

must be inconceivably wild.[65]

At Valparaiso Darwin was able to admire the volcano of Aconcagua – 23,000 feet in height. On 14 August he

> set out on a riding excursion for the purpose of geologizing the basal parts of the Andes [and declared] The proofs of the elevation of this whole line of coast are unequivocal: at the height of a few hundred feet old-looking shells are numerous, and I found some at 1300 feet.[66]

Darwin realized that the presence of sea shells in the foothills of the Andes mountains indicated that these foothills had once been at sea level. In that same month he told Caroline that

> The ultimate destination of *all* my collections will of course be to wherever they may be of most service to Natural History. But caeteris paribus [other things being equal] the British Museum [London, devoted to human history and culture], has the first claims, owing to my being on board a King's Ship.[67]

Darwin arrived at Santiago, capital of Chile, on 27 August. The following day he gleefully told Captain FitzRoy, 'Altogether I am delighted with the Country of Chile … .'[68] However, on 19 September he remarked, 'During the day I felt very unwell, and from that time until the end of October did not recover'.[69]

> 24 September 1834. Our course was now directed towards Valparaiso [Chile], which with great difficulty I reached on the 27th, and was there confined to my bed till the end of October.[70]

Darwin summed up the nature of HMS *Beagle*'s captain thus

> Fitz-Roy's character was a singular one, with very many noble features: he was devoted to his duty, generous to a fault, bold, determined, and indomitably energetic, and an ardent friend to all under his sway. He would undertake any sort of trouble to assist those whom he thought deserved assistance.[71]

However, there was a negative side to the captain, for, said Darwin, his

> temper was a most unfortunate one. It was usually worst in the early morning, and with his eagle eye he could generally detect something amiss about the ship, and was then unsparing in his blame. He was very kind to me, but was a man very difficult to live with on the intimate terms which necessarily

> followed from our messing by ourselves in the same cabin.[72]

This statement by Darwin is a tribute to his ability to 'get along' with his fellow human beings, even in the most trying of circumstances, bearing in mind the cramped confines of the ship.

Writing from Valparaiso on 8 November, Darwin informed Catherine that FitzRoy had been 'unwell', largely through overwork.

> This was accompanied by a morbid depression of spirits, & a loss of all decision & resolution. The captain was afraid that his mind was becoming deranged (being aware of his hereditary predisposition).[73]

Perhaps it was his uncle, politician Robert Stewart, Viscount Castlereagh, who had committed suicide in 1822, that Captain FitzRoy had in mind when he expressed these fears.

> 10 November 1834. The *Beagle* sailed from Valparaiso to the south, for the purpose of surveying the southern part of Chile ...[74]

In a long letter, written between July and November 1834, Darwin informed Henslow that his notes, which were 'becoming bulky', comprised

> about 600 small quarto pages full; about half of this is Geology, the other imperfect descriptions of animals: with the latter I make it a rule only to describe those parts, or facts, which cannot be seen, in specimens [preserved] in spirits.

Also, Darwin was anxious for Henslow to confirm the safe arrival of the head (i.e. skull) of a megatherium (giant sloth), which he had sent to him from Buenos Aires.[75]

> 15 January 1835 ... we sailed from Low's Harbour, and three days afterwards anchored a second time in the Bay of S. Carlos in [the island of] Chiloe [off the coast of southern Chile].[76]

Whereupon the indefatigable Darwin set off on a 'twelve-league' [approximately thirty-six-mile] excursion across the island.[77] He had now, seemingly, fully recovered his health and strength.

> Catherine wrote to Darwin on 28 January to say that their father Robert – 'Papa' wishes to urge you to think of leaving the *Beagle*, and returning home, and to

> take warning by this one serious illness … [Darwin believed that he had fallen ill as a consequence of drinking 'bad wine'. He] desires me to beg you to recollect that it will soon be four years since you left us, which surely is a long portion of your life to give up to Natural History. – If you wait till the *Beagle* returns home, it will be as many years again; the time of its voyage goes on lengthening & lengthening every time we hear of it; we are quite in despair about it.[78]

Fortunately for posterity, Darwin ignored his father's entreaties, HMS *Beagle* sailed from Chiloe on 4 February and reached Valdivia (Chile) on the night of the 8th.[79] It was here, twelve days later, that Darwin experienced 'the most severe earthquake' to have occurred in the living memory of the inhabitants.[80] The following month, Darwin described 'the great Earthquake of 20th of February' to Caroline.

> I suppose it certainly is the worst ever experienced in Chili [Chile]. It is no use attempting to describe the ruins – it is the most awful spectacle I ever beheld. The town of [Concepción] is now nothing more than piles & lines of bricks, tiles & timbers – it is absolutely true there is not one *house* left habitable … The force of the shock must have been immense, the ground is traversed by rents, the solid rocks are shivered, solid buttresses 6-10 feet thick are broken into fragments like biscuit.[81]

According to Captain FitzRoy

> At ten in the morning of 20 February, very large flights of sea-fowl were noticed, passing over the city of Concepción, from the sea-coast, towards the interior: and some surprise was excited by so unusual and simultaneous a change in the habits of those birds, no signs of an approaching storm being visible.
>
> About eleven, the southerly breeze freshened up as usual – the sky was clear, and almost cloudless. At forty minutes after eleven the shock of an earthquake was felt, slightly at first, but increasing rapidly. During the first half minute many persons remained in their houses; but then the convulsive movements were so strong, that the alarm became general, and they all rushed into open spaces for safety. The horrid motion increased: people could hardly stand; buildings waved and tottered – suddenly an awful overpowering shock caused universal destruction – and in less than 6 seconds the city was in ruins.[82]

The earthquake was followed by what would now be called a tsunami, described by Darwin thus

> Shortly after the shock, a great wave was seen from the distance of three or

> four miles, approaching the middle of the bay with a smooth outline; along the shore it tore up cottages and trees as it swept onwards with irresistible force.[83]

And by Captain FitzRoy as follows

> an enormous wave was seen forcing its way through the western passage which separates Quirquina Island from the mainland. This terrific swell passed rapidly along the western side of the Bay of Concepción, sweeping the steep shores of every thing movable within thirty feet from the high-water mark.[84]

Darwin observed that 'in almost every earthquake the neighbouring waters of the sea are said to have been greatly agitated', and he pondered over why this should be so.[85] He also observed that after the earthquake, 'the land round the Bay of Concepción was raised two or three feet [and] at the island of S. Maria (about thirty miles distant) the elevation was greater …'.[86] It was therefore his opinion that 'successive small uprisings, such as that which had accompanied or caused the earthquake of this year …' had caused the 'great elevation'. This was the explanation, for example, for sea shells being 'found at the height of 1300 feet' at Valparaiso (Chile).[87]

> 7 March 1835. We stayed three days at Concepción, and then sailed for Valparaiso.[88]

From here, on 11 March, Darwin set out on an expedition to cross the Cordillera (mountain ranges)[89]

> Our first day's journey was fourteen leagues [forty-two miles] to Estacado [Chile], and the second seventeen to [Luján] near Mendoza [Argentina, situated 200 miles east of Valparaiso]. We crossed the Luxan, which is a river of considerable size …[90]
>
> At night I experienced an attack (for it deserves no less a name) of the *Benchuca*, a species of Reduvius, the great black bug of the Pampas.[91]

This incident, which will be discussed shortly, would prove to be of great significance, as far as Darwin's health was concerned

> 27 March 1835. We rode on to Mendoza.[92] On the 10th we reached Santiago … . My excursion only cost me twenty-four days, and never did I more deeply enjoy an equal space of time.[93]

On 18 April, writing from Valparaiso, Darwin told Henslow

> I have just returned from Mendosa, having crossed the Cordilleras [mountain

> ranges] by [way of] two passes. This trip has added much to my knowledge of the geology of the country.[94]

Five days later, in a letter to his sister Susan, he elaborated further.

> Beside understanding to a certain extent, the description & manner of the force, which has elevated this great line of mountains, I can clearly demonstrate, that one part of the double line is of a age long posterior to [i.e. much older than] the other. In the more ancient line, which is the true chain of the Andes. I can describe the sort & order of the rocks which compose it.
>
> What is of much greater consequence, I have procured fossil shells (from an elevation of 12000 ft). I think an examination of these will give an approximate age to these mountains as compared to the Strata of Europe: In the other line of the Cordilleras there is a strong presumption (in my own mind conviction) that the enormous mass of mountains, the peaks of which rise to 13 & 14000 ft are so very modern as to be contemporaneous with the plains of Patagonia (or about [contemporaneous] with [the] *upper* strata of [the] Isle of Wight): If this result shall be considered as proved it is a very important fact in the theory of the formation of the world.[95]

(It was Danish Catholic bishop and scientist Nicholas Steno, 1638–86, who was instrumental in defining the guiding principles of Stratigraphy – that branch of geology which is concerned with the order and relative position of strata and their relationship to the geological time scale.[96] He argued that rock layers (or strata) are laid down in succession, and that each represents a 'slice' of time. He also formulated the law of superposition, which states that any given stratum is probably older than those above it and younger than those below it. This, of course, is a generalization, and takes no account of distortions to the Earth's crust, which have taken place over time and continue to do so.)

> 27 April 1835. I set out on a journey to Coquimbo [Chile], and thence through Guasco to Copiapo where Captain Fitz Roy kindly offered to pick me up in the *Beagle*. The distance in a straight line along the shore northward is only 420 miles; but my mode of travelling made it a very long journey.

The journey was, as usual, undertaken on horseback, during the course of which Darwin and his party cooked their own meals and slept in the open air.[97]

As if to demonstrate the ever-present danger of disease, FitzRoy describes how, on 25 May, near a village called Quiapo [Chile]

> When I stripped to bathe … I found myself so covered from head to foot with

> flea-bites, that I seemed to have a violent rash, or the scarlet fever.[98]

Meanwhile, continued Darwin

> 12 July 1835. We anchored off the port of Iquique ..., on the coast of Peru.[99]

> 19 July 1845. We anchored in the Bay of Callao, the seaport of Lima, the capital of Peru.[100]

To Caroline in that same month, Darwin wrote from Lima to say, in eager anticipation

> I am very anxious for the Galapagos Islands, – I think both the Geology & Zoology cannot fail to be very interesting.[101]

In August Darwin wrote to Fox to congratulate him on his recent marriage.

> You are a true Christian & return good for evil.- to send 2 letters to so a bad a Correspondent as I have been. God bless you for writing so kindly & affectionately ...
>
> The voyage is terribly long. I do so earnestly desire to return, yet I dare hardly look forward to the future, for I do not know what will become of me. Your situation is above envy; I do not venture even to frame such happy vision. To a person fit to take the office, the life of a Clergyman is a type of all that is respectable & happy: & if he is a Naturalist & has the 'Diamond Beetle' [*Curculio imperialis* – a Brazilian beetle notable for its sparkling wing-cases – 'elytra'], ave Maria ... .[102]

This letter to Fox begs the question: was Darwin debating in his mind whether he himself was 'fit' – i.e. suitable and suited, and whether he possessed the necessary conviction – which Fox undoubtedly did, to become a clergyman, as his father Robert, wished for him to be?

On the 15th of that month, Darwin wrote from Lima to Henry Stephen Fox (British minister-plenipotentiary and envoy-extraordinary at Rio de Janeiro, and amateur geologist) to say

> I hope now to be able to give some sort of outline of the superposition of the strata & the structure of the mountains of [Chile]. It is very certain that the very idea of the Cordilleras being composed solely of Volcanic rocks is quite incorrect. [Also] There is one point in the Geology of S. America, in which I am much interested, it is the recent elevation of the land. That such has taken

> place & to a considerable amount on this coast I have abundant proofs. [He then enquires] Have you ever noticed on land elevated from 30 to 200 ft above the sea, any *large* beds of marine shells, & which did not appear carried there by man?[103]

In September 1835 to Alexander Burns Usborne, naval officer aboard *Beagle*, Darwin declared of shells which had once lain on the sea bed

> I have frequently found such shells on the coast of Chili [sic] at a height from 20 – 400 ft. The oftener you can observe & record this class of facts, in different places, so much the better; for the evidence respecting the rise of land becomes cumulative.[104]

HMS *Beagle* arrived at the Galapagos Islands on 15 September 1835.

Chapter 9

The Galapagos

Darwin described the Galapagos, with which his name would forever thereafter be associated, thus:

> This archipelago consists of ten principal islands of which five exceed the others in size. They are situated under the equator [in fact, the Galapagos archipelago is bisected by the equator] between five and six hundred miles westward of the coast of America. They are all formed of volcanic rocks … .[1]
>
> Most of the organic [living] productions are aboriginal creations found nowhere else; there is even a difference between the inhabitants of the different islands; yet all show a marked relationship with those of America, though separated from that continent by an open space of ocean between 500 and 600 miles in width. The archipelago is a little world within itself, or rather a satellite attached to America … .[2]

Here, Darwin collected and studied twenty-six species of land birds, eleven species of waders and water birds, together with reptiles, sea turtles, a lizard, a snake, and tortoises. He also collected fish and sea shells and declared that of the two hundred and twenty-five species of flowering plants which he had discovered in the Galapagos '100 are new species and are probably confined to this archipelago'.[3] But, said he,

> The most remarkable feature in the natural history of this archipelago by far … is that the different islands to a considerable extent are inhabited by a different set of beings. My attention was first called to this fact by the Vice-governor, Mr. [Nicolas] Lawson, declaring that the tortoises differed from [i.e. on] the different islands, and that he could with certainty tell from which island any one was brought. I did not for some time pay sufficient attention to this statement, and I had already partially mingled together the collections from two of the islands. I never dreamed that islands, about 50 or 60 miles apart, and most of them in sight of each other, formed of precisely the same rocks, placed under a

> quite similar climate, rising to a nearly equal height, would have been differently tenanted; but we shall soon see that this is the case.[4]
>
> For example, Captain [David] Porter [of the US Navy] has described those from Charles [Island] and from the nearest island to it, namely, Hood Island, as having their shells in front thick and turned up like a Spanish saddle, whilst the tortoises from James Island are rounder, blacker and have a better taste when cooked.[5]

Darwin himself remarked, in respect of the mocking thrushes, that

> to my astonishment, I discovered that all those from Charles Island belonged to one species (Mimus trifasciatus), all from Albemarle Island to M. parvulus, and all from James and Chatham islands belonged to M. melanotis.[6]
>
> But it is the circumstance [i.e. fact] that several of the islands possess their own species of the tortoise, mocking-thrush, finches, and numerous plants, these species having the same general habits, occupying analogous situations, and obviously filling the same place in the natural economy of this archipelago, that strikes me with wonder.[7]

Instead of dumbly and unquestioningly accepting these curious facts, Darwin would mull them over in his mind, until one day, in a dazzling flash of inspiration, an explanation for them would be forthcoming. Meanwhile, he continued his narrative thus

> 22 October 1835. The survey of the Galapagos Archipelago being concluded, we steered towards Tahiti [in the South Pacific] and commenced our long passage of 3220 miles.[8]
>
> 15 November 1835. At daylight, Tahiti ... was in view.[9]
>
> 17 November 1835. After breakfast I went on shore and ascended the nearest slope to a height of between two and three thousand feet.[10]

At Tahiti, on 21 November, Captain FitzRoy was told by one of the native inhabitants that

> the island was not so healthy as in former times; and they [the inhabitants] had caught diseases, in those days unknown. Asking who brought this or that disease, he imputed the worst to the ships which came after Cook's first visit and left men upon the island until their return the following year. Curvature of the spine, or a hump-back, never appeared until after Cook's visits, and as he had a hump-backed man in his ship, they attribute that deformity to him.[11]

(This was a reference to British explorer Captain James Cook's 1769 expedition to the island. One cause of spinal curvature is Potts' disease – infection of the spinal vertebrae with tuberculosis.)

> 26 November 1835. In the evening, with a gentle land breeze, a course was steered for New Zealand ...[12]

> 19 December 1835. In the evening we saw in the distance New Zealand.[13]

> 21 December 1835. Early in the morning we entered the Bay of Islands [North Island, New Zealand] ...[14]

> 30 December 1835. In the afternoon we stood out of the Bay of Islands, on our course to Sydney [Australia].[15]

> 12 January 1836. Early in the morning a light air carried us towards the entrance of Port Jackson [Sydney Harbour].[16]

At Sydney, Darwin and FitzRoy were re-united temporarily with Conrad Martens, who had established a reputation as an artist of note. From him, a number of paintings were commissioned of *Beagle*'s visits to Tierra del Fuego and the Pacific.

With the native inhabitants of Australia, and those of other continents in mind, Darwin declared

> Wherever the European has trod, death seems to pursue the aboriginal [meaning, in this context, human beings indigenous to the land from the earliest times]. We now look to the wide extent of Polynesia, the Cape of Good Hope, and Australia, and we find the same result. Nor is it the white man alone that acts [as] the destroyer ... . The varieties of man seem to act on each other in the same way as different species of animals – the stronger always extirpating the weaker.[17]

> 30 January 1836. The *Beagle* sailed for Hobart Town in Van Diemen's Land [Tasmania].[18]

> 7 February 1836. The *Beagle* sailed from Tasmania ...[19]

> 1 April 1836. We arrived in view of the Keelling or Cocos Islands, situated in the Indian Ocean ...[20]

> 12 April 1836. In the morning we stood out of the lagoon on our passage to the Isle of France [Mauritius, in the Indian Ocean].[21]

Here, Darwin took the opportunity to study in great detail the various types of coral reef, and expound his theories as to how they originated.[22]

> 29 April 1836. In the morning we passed round the northern end of Mauritius … .[23]

> 9 May 1836. We sailed from Port Louis [capital of Mauritius] and, calling at the Cape of Good Hope, on the 8th July we arrived off St Helena [island in the South Atlantic,[24] where Darwin declared] I so much enjoyed my rambles among the rocks and mountains …[25]

On 19 July, HMS *Beagle* reached the island of Ascención in the South Atlantic, where Darwin 'ascended Green Hill, 2840 feet high'.[26]

> On leaving Ascencion we sailed for Bahia [Salvador], on the coast of Brazil [which *Beagle* had first visited in February 1832], in order to complete the chronometrical measurement of the world. We arrived there on August 1st and stayed four days, during which I took several long walks.[27]

> 19 August 1836. …we finally left the shores of Brazil … . I thank God that I shall never again visit a slave country [i.e. a country where people were kept in slavery]. I will not even allude to the many heart-sickening atrocities I authentically heard of … .[28]

If Darwin had doubts about the Christian doctrine, they were certainly not apparent when he wrote these words of optimism in relation to the southern hemisphere.

> From seeing the present state, it is impossible not to look forward with high expectations to the future progress of nearly an entire hemisphere [i.e. the southern hemisphere]. The mark of improvement, consequent on the introduction of Christianity throughout the South Sea, probably stands by itself in the records of history.[29]

> 21 August 1836. … we anchored for the second time at Porto Praya in the Cape de Verd [e] [North Atlantic island, off the coast of Senegal], where we stayed six days. On the 2nd of October we made the shore of England; and at Falmouth I left the *Beagle*, having lived on board the good little vessel nearly five years.[30]

Darwin's thoughts at this time can scarcely be imagined: relief to have returned home safely; satisfaction at having seen the task through and fulfilled his mission; wonder at the wonderful sights he had seen; pleasure at the prospect of discussing the specimens which he had collected with his scientific colleagues, and, of course, joy at the prospect of being reunited with his beloved family.

Chapter 10

Home at Last

Following his return to England, Darwin, on 13 September 1832

> settled in lodgings in Cambridge … where all my collections [of specimens] were under the care of Henslow. I stayed here three months, and got my minerals and rocks examined with the aid of Professor Miller [William Hallowes Miller, Cambridge University's Professor of Mineralogy].[1]

The voyage in retrospect

The voyage of the *Beagle*, said Darwin,

> has been by far the most important event in my life, and has determined my whole career … I was led to attend closely to several branches of natural history, and thus my powers of observation were improved …

Now was the time to assimilate all that he had seen, learned and discovered, and to bring his analytical mind to bear upon what he identified as unexplained phenomena. However, said he

> The investigation of the geology of all the places visited was far more important, as reasoning here comes into play. On first examining a new district, nothing can appear more hopeless than the chaos of rocks; but by recording the stratification and nature of the rocks and fossils at many points, always reasoning and predicting what will be found elsewhere, light soon begins to dawn on the district, and the structure of the whole becomes more or less intelligible.[2]

(The rocks in which fossils occur are of the 'sedimentary' type – i.e. which have formed from sediment deposited by water or air, rather than a), 'igneous' – which, when it occurs above ground, is rock which has solidified from lava or magma, as a result of volcanic processes; or b), 'metamorphic' – denoting rock that has undergone transformation by heat, pressure, or other natural agencies, e.g., in the folding of strata or the nearby intrusion of igneous rocks.)[3]

Simply collecting and admiring rocks and rock formations was not sufficient for Darwin, whose satisfaction came from bringing his thought processes to bear in order to ascertain the nature of the rocks, and how such formations came into existence.

> Another of my occupations was collecting animals of all classes, briefly describing and roughly dissecting many of the marine ones … .[4]
>
> I also reflect with high satisfaction on some of my scientific work, such as solving the problem of the coral islands, and making out the geological structure of certain islands, for instance, St Helena.

A coral, or coral polyp, is an invertebrate marine animal, tubular and sac-like in shape; a few millimetres in diameter and a few centimetres in length. At one end is a mouth surrounded by a ring of tentacles and at the other, a base which is attached to the underlying stratum (e.g. rock). Polyps live in colonies and thrive in shallow, tropical waters. Near to their base the coral secretes calcium carbonate, which forms a hard exoskeleton (external covering).

Darwin was intent on finding out how coral reefs and atolls are created (an atoll being a ring-shaped reef, island, or chain of islands formed of coral[5]).

> No other work of mine was begun in so deductive a spirit as this, for the whole theory was thought out on the west coast of South America, before I had seen a true coral reef. I had therefore only to verify and extend my views by a careful examination of living reefs.[6]

As for St Helena, Darwin deduced correctly, that 'The whole island is of volcanic origin …[7]'. But, equally intriguing, was something which Darwin had discovered from his reading

> accounts are given of a series of volcanic phenomena – earthquakes – troubled water – floating scoriae [volcanic rock] and columns of smoke – which have been observed at intervals since the middle of the last century, in a space of open sea between longitudes 20° and 22° west, about half a degree south of the equator. These facts seem to show that an island or an archipelago is in process of formation in the middle of the Atlantic: a line joining St Helena and Ascension, [if] prolonged intersects this slowly nascent focus of volcanic action.[8]

Darwin was referring to a geological feature, now known as the Mid-Atlantic Ridge (southern portion), which marks the boundary of two tectonic plates. Here, as the continents of South America and Africa drift apart, magma from the Earth's mantle wells up through the crack between the plates, replenishing the Earth's crust and sometimes creating volcanoes high enough to protrude above the surface of the ocean.

> Nor must I pass over the discovery of the singular relations of the animals and plants inhabiting the several islands of the Galapagos archipelago, and all of them to the inhabitants of South America.[9]

The puzzle as to why different varieties of the same species occurred on the various islands of the Galapagos nagged away in Darwin's mind as something for which there *must* be an explanation. It was a loose end, and he could not bear loose ends!

> I was also ambitious to take a fair place among scientific men … . [He admitted, however, that] I have never turned one inch out of my course to gain fame.[10]

* * *

On 5 October 1836 Darwin wrote to Josiah Wedgwood (II) to say, 'I am most anxious once again to see Maer [Hall, home of the Wedgwood family] and all its inhabitants, so that in the course of two or three weeks I hope in person to thank you, as being my First Lord of the Admiralty.' This was an acknowledgement by Darwin, of the fact that Josiah's influence had 'induced his father to consent to his joining the *Beagle*'.[11]

FitzRoy wrote to Darwin on 20 October as follows, addressing him affectionately as 'Philos' (after the Hellenistic Jewish philosopher Philo Judaeus).

> Dearest Philos …
> Believe it, or not, – the news is *true* – I am going to be *married*!!!!!! To Mary O'Brien … . All is settled and we shall be married in December. Yours most sincerely, Robt FitzRoy.[12]

In his letter to Henslow of 30/31 October Darwin declared:

> I have not made much progress with the great men [of science, whom he had tried to interest in the collections of specimens which he had brought back from South America]. I find, as you told me, that they are overwhelmed with their own business. [However] Mr Lyell has entered in the *most* goodnatured manner, & almost without being asked, into all my plans.

Sir Charles Lyell was a Scottish barrister. He was also President of the Geological

Society and a man whom Darwin greatly respected, and about whom he said, 'The science of Geology is enormously indebted to Lyell – more so, as I believe, than to any other man who ever lived.[13] He tells me, however, the same story, namely that I must do all myself. Mr Owen seems anxious to dissect some of the animals in spirits, & besides these two I have scarcely met anyone who seems to wish to possess any of my specimens.'

Richard Owen, a comparative anatomist, was Hunterian Professor in the Royal College of Surgeons. Continued Darwin to Henslow:

> As far as I can yet see, my best plan will be to spend several months in Cambridge, & then, when by your assistance, I know on what grounds I stand, to emigrate to London, where I can complete my geology, & *try to push on* [with] *the Zoology*.[14]

From this, it may be deduced a), that Darwin is disappointed in the lack of interest shown in the specimens which he had collected so painstakingly and laboriously during the *Beagle* voyage and b), that as far as he was concerned, all thoughts of becoming a clergyman in the Church of England had vanished.

Astonishingly, mislaid artefacts collected by Darwin during the course of the *Beagle* expedition, were only rediscovered in April 2011 by Dr Howard Falcon-Lang, a palaeobotanist at Royal Holloway, University of London, when, in a 'gloomy corner' of the headquarters of the British Geological Survey, Nottingham, he had 'stumbled upon … an old wooden cabinet' which contained glass slides on which had been placed for examination specimens of fossils. One of the first slides Falcon-Lang had come across was, to his astonishment, labelled 'C. Darwin Esq'.

It transpired that the collection consisted of 314 slides of specimens, collected not only by Darwin but also by botanist and explorer Joseph Dalton Hooker, John Stevens Henslow and others. One, for example, depicted a section of fossilized wood, 40 million years old and collected by Darwin in 1834 from Chiloe Island, South America during the voyage of the *Beagle*.[15]

In that same month of October, Emma Wedgwood, Darwin's cousin (and daughter of Josiah Wedgwood II), who was now aged twenty-eight, declared

> We are getting impatient for Charles's arrival. We all ought to get up a little knowledge for him. I have taken to no deeper study than Capt Head's gallop which I have never read before.

The book to which Emma referred was entitled *Rough Notes taken during some Rapid Journeys Across the Pampas and among the Andes* by Sir Francis Bond Head, Lieutenant-Governor of Upper Canada and known as 'galloping Head', published in 1828.[16]

Was Emma's desire to be reunited with Darwin simply occasioned by a sense of family love and loyalty, or was there a deeper reason? Time would tell.

On 21 November, following Darwin's visit to Maer, Emma declared, 'We enjoyed Charles's visit uncommonly. [He] talked away most pleasantly all the time; we plied him with questions without any mercy.[17]'

Emma Wedgwood

Emma Wedgwood was to play an increasingly important role in Darwin's life. She was born on 2 May 1808, the youngest of the eight surviving children of Josiah Wedgwood (II) and his wife Elizabeth, née Allen, of Maer Hall, Staffordshire. Emma was, therefore, almost a year older than Darwin (who, as already mentioned, was her cousin: Darwin's father Robert, having married Josiah Wedgwood (I)'s daughter Susannah, on 18 April 1796).

As might be expected of two families which were so closely related, there was frequent interaction between them. For example, when, in summer 1823, the Wedgwoods paid a visit to the seaside resort of Scarborough, Yorkshire, they were joined by Darwin's sister Susan.[18]

Emma kept a diary that reveals that her reading matter was largely of a religious nature. For example, she lists the books which she read in the year 1824. They included *Moral Philosophy* and *Horae Paulinae, or the Truth of the Scripture History of St Paul*, by the Reverend William Paley; *Isaiah: A New Translation*, by Bishop Robert Lowth, and *Sermons on the Efficacy of Prayer and Intercession* by the Reverend Samuel Ogden.[19]

When Emma left school she taught at the Sunday school (as did other members of her family), which was attended by sixty children and held in the laundry room of Maer Hall.[20] In March 1827, Emma and her family visited Geneva. That autumn Darwin's sisters Susan and Catherine, visited Maer for a month, where 'Emma Wedgwood, now nineteen, was leading a happy, girlish life, taking what parties, balls and archery meetings came in her way'.[21]

When Emma's beloved older sister Francis ('Fanny'), died on 20 August 1832 at the age of twenty-six, Emma recorded her feelings.

> Oh, Lord, help me to become more like her, and grant that I may join her with Thee never to part again.[22]

* * *

Lyell wrote an admiring letter to Darwin in February 1837 referring to

> your new Llama [the South American guanaco, or wild llama, which Darwin had described], Armadillos, gigantic rodents, & other glorious additions to the

> Menagerie of that new continent .... [Furthermore] Your lines of Elevation & subsidence will deservedly get you as great a name as de Beaumont's parallel Elevations, & yours are true, which is more than can be said of his.[23]

This was a reference to French geologist Jean-Baptiste de Beaumont and his theory of the origin of mountain ranges.

On 7 March, recorded Darwin,

> I took lodgings in Great Marlborough Street in London, and remained there for nearly two years, until I was married. During these two years I finished my Journal, read several papers before the Geological Society, began preparing the MS [manuscript] for my *Geological Observations*, and arranged for the publication of the *Zoology of the Voyage of the Beagle*. In July I opened my first note-book for facts in relation to the *Origin of Species*, about which I had long reflected, and never ceased working for the next twenty years.[24]

By May/June, Darwin was exercising his mind as to why it was that in

> so many cases where ships with all their crew in good health have yet caused strange contagious disorders at [British East India Company's Naval] Station Islands in the Pacific ... [and how it was that the] first mingling ... of the European race with the natives of distant climes always produces disease.[25]

Darwin wrote to William Whewell, Professor of Mineralogy and Moral Theology at Cambridge and President of the Geological Society on 18 June to say, 'I have been much interested on the subject of earthquake waves [i.e. tsunamis.]'.[26] However, the letter contained an item of far greater significance.

> My first note-book was opened in July 1837. I worked on true Baconian principles [moving from specific observations to broader generalisations and theories, as practised by philosopher Francis Bacon, 1561–1626], and without any theory collected facts on a wholesale scale, more especially with respect to domesticated productions, [animals and plants which had been selectively bred] by printed enquiries [to fellow enthusiasts], by conversation with skilful breeders and gardeners, and by extensive reading. I soon perceived that selection was the keystone of man's success in making useful races of animals and plants. But how selection could be applied to organisms living in a state of nature remained for some time a mystery to me.[27]

By recognizing the possible significance of 'selection' in respect of the natural world, Darwin had taken another step towards formulating his great theory.

Darwin wrote to John Richardson, surgeon, explorer, and naturalist on 24 July with questions about vegetation in cold climates.

> My object in these questions, is to be enabled to compare the mere *quantity* of vegetation, in parts of South America, where large animals formerly did live, and likewise in Africa where large animals are now living, with the quantity growing in climates far north, and extremely cold.[28]

During his time in London, Darwin, who was ever-inquisitive and whose mind ranged freely across the entire panoply of the natural world

> read before the Geological Society papers on the 'Formation by the Agency of Earth-worms of Mould' (1838), 'Earthquakes' (1840), and the 'Erratic Boulders of South America' (1842).[29]

In a letter to Susan in April 1838 Darwin described a visit he had made to the Zoological Society (the London Zoo) where he saw

> the [Orang-utan] in great perfection: the keeper showed her an apple, but would not give it [to] her, whereupon she threw herself on her back, kicked & cried, precisely like a naughty child. She then looked very sulky & after two or three fits of [passion] the keeper said, 'Jenny if you will stop bawling & be a good girl, I will give you the apple. She certainly understood every word of this, &, though like a child, she had great work [i.e. was obliged to make a great effort] to stop whining, she at last succeeded, & then got the apple, with which she jumped into an arm chair & began eating it, with the most contented countenance imaginable.[30]

Darwin was elected to the Athenaeum on 21 June. This was a club founded in 1824 for men of distinction in the fields of medicine, law, the arts, etc., and also for high-ranking clergymen. To Lyell, who had supported him in his candidature, he declared

> I am full of admiration for the Athenaeum; one meets so many people there, that one likes to see … Your helping me into the Athenaeum has not been thrown away, & I enjoy it the more because I fully expected to detest it.[31]

From August 1838 until December 1881 Darwin kept a *Journal* in which he 'recorded the periods he was away from home, the progress and publication of his work and important events in his family life'.[32] That summer, in the typically analytical manner of the trained scientist, he listed what, to his mind, were the pros and cons of marriage.

> *Marry*
> Children – (if it Please God) Constant companion (& friend in old age) who will feel interested in one, object to be beloved & played with. better than a dog anyhow. Home & someone to take care of house. Charms of music & female chit-chat. These things good for one's health. *but terrible loss of time*.
> My God, it is intolerable to think of spending one's whole life, like a neuter bee, working, working & nothing after all. No, no won't do. Imagine living all one's days solitarily in smoky dirty London House. Only picture to yourself a nice soft wife on a sofa with good fire, & books & music perhaps – Compare this vision with the dingy reality of Grt Marlbro' Street.
>
> *Not marry*
> Freedom to go where one liked – choice of Society & *little of it.* Conversation of clever men at clubs– not forced to visit relatives, & to bend in every trifle. to have the expense & anxiety of children – perhaps quarrelling – *Loss of time* [it was no coincidence that these two words were highlighted by him, workaholic that he was]. cannot read in the Evenings – fatness & idleness – Anxiety & responsibility – less money for books &c – if [I have] many children [to be] forced to gain one's bread [i.e. work in a conventional way in order to earn a living]. (But then it is very bad for one's health to work too much).
> Perhaps my wife won't like London; then the sentence is banishment & degradation into indolent, idle fool … .

As the pros outweighed the cons, the conclusion he reaches, therefore, is 'Marry – Mary [sic] – Marry. QED.' Finally, 'It being necessary to Marry', he poses the question, 'When? Soon or Late'.[33]

The fact that Darwin does not include 'To whom?' in the question, indicates that he has already made up his mind as to the subject of his love and affection – i.e. Emma Wedgwood.

Were it not for the words 'object to be beloved', written in relation to a prospective spouse, one might be forgiven for believing that Darwin regarded marriage merely as a cold, calculated transaction, based on the premise 'What is there in this for me?' But what did Darwin himself have to offer? Was this scientist with an analytical mind capable of fulfilling the role of husband, and subsequently perhaps, parent, with all that this entailed, including the *giving* of love and emotional support? Only time would tell.

In his autobiography, Darwin made the following, highly significant entry

In October 1838, that is, fifteen months after I had begun my systematic enquiry [into 'how selection could be applied to organisms living in a state of nature'], I happened to read for amusement [Thomas R.] Malthus [economist and clergyman] on *Population* ...[34]

Chapter 11

Thomas Robert Malthus

Thomas Malthus, clergyman and economist, was born in 1766. In 1788 he graduated from Jesus College, Cambridge, and immediately took Holy Orders. Of his belief in Christianity there is no doubt. Said he

> Evil exists in the world not to create despair but activity. We are not patiently to submit to it, but to exert ourselves to avoid it. [And] the more wisely he [the individual] directs his efforts … the more he will probably improve and exalt his own mind and the more completely does he appear to fulfil the will of his Creator.[1]

In his magnum opus, *An Essay on the Principle of Population* (published in 1798), he wrote

> I think I may fairly make two postulata.
> First, that food is necessary to the existence of man.
> Secondly, that the passion between the sexes is necessary and will remain nearly in its present state.[2]
>
> Population, when unchecked, increases in a geometrical ratio. Subsistence ['the action or fact of maintaining or supporting oneself at a minimum level'] increases in an only arithmetical ratio. A slight acquaintance with numbers will shew the immensity of the first power in comparison of [with] the second.
>
> By that law of our nature which makes food necessary to the life of man, the effects of these two unequal powers must be kept equal.
>
> This implies a strong and constantly operating check on population from [i.e. because of] the difficulty of subsistence [i.e. in the face of a food supply which is struggling to keep pace].[3]

'So diversified are the natural objects around us', wrote Malthus, and 'so many instances of mighty power [i.e. the power of the Deity] daily offer themselves to our view'. Such seemingly 'trifling … little bits of matter', as 'a grain of wheat, and an acorn' were

> possessed [of] such curious powers of selection, combination, arrangement, and almost of creation, that upon being put into the ground, they would choose, amongst all the dirt and moisture that surrounded them, those parts which best suited their purpose, that they would collect and arrange these parts with wonderful taste, judgement, and execution, and would rise up into beautiful forms, scarcely in any respect analogous to the little bits of matter which were first placed in the earth.[4]
>
> The powers of selection, combination, and transmutation, which every seed shews, are truly miraculous.

What did Malthus mean by the terms 'selection' and 'transmutation'? To him, 'selection' meant the ability of seeds, by some means which he did not specify, to transform themselves – 'transmute' - so as to make the very best of the environment in which they found themselves, and thus 'rise up into beautiful forms'.[5]

> Many vessels will necessarily come out of this great [metaphorical] furnace in wrong shape. These will be broken and thrown away as useless; while those vessels whose forms are full of truth, grace, and loveliness will be wafted into happier situations, nearer the presence of the mighty maker.[6]

Limitations to food production

In another work entitled *A Summary View of the Principle of Population* (first published in 1830), Malthus wrote of how 'The scarcity of [fertile] land [and] the great natural barrenness of a very large part of the surface of the earth,' set limits to the amount of food which may be produced.[7]

Factors which favoured an increase in population

Malthus listed these factors as freedom 'from scarcities and epidemics ..., the healthiness of the country [and] the introduction of vaccination'.[8]

Ireland as an example

Malthus cites Ireland where, between the years 1695 and 1821, the population had increased from 1,034,000 to 6,801,827, which was equivalent to a doubling 'in about forty-five years'. And he points out that this increase had caused 'great distress among the labouring classes of society' and had led to 'the practice of frequent and considerable emigration'.[9]

The problem in a nutshell

'It may be safely asserted ...' said Malthus, 'that population, when unchecked, increases in a geometrical progression of such a nature as to double itself every twenty-five years'.[10] In contrast, whereas

> each farm in the well-peopled countries of Europe might allow of one or even two doublings [of food production], without much distress … the absolute impossibility of going on at the same rate is too glaring to escape the most careless thinker.[11]

A way forward: the perils of failure

> the only mode of keeping population on a level with the means of subsistence which is perfectly consistent with virtue and happiness … [was] moral restraint … [which] may be defined to be, abstinence from marriage, either for a time or permanently … . All other checks … resolve themselves into some form of vice or misery [such as] war and violent diseases … plagues, famines, and mortal epidemics …[12]

However, Malthus admits that 'Prudence [i.e. sexual abstinence] cannot be enforced by laws, without a great violation of natural liberty, and a great risk of producing more evil than good'.[13]

In conclusion, Malthus declared that:

> The evils arising from the principle of population [outstripping the food supply] as are of exactly the same kind as the evils arising from the excessive or irregular gratification of the human passions in general, and may equally be avoided by moral restraint.[14]
>
> It will be acknowledged that in a state of probation [the process or period of testing or observing the character or abilities of a person in a certain role[15]], those laws which seem best to accord with the views of a benevolent Creator which, while they furnish the difficulties and temptations which form the essence of such a state, are of such a nature as to reward those who overcome them, with happiness in this life as well as in the next.[16]

Malthus, a clergyman, regarded his advocacy of moral restraint as a means of limiting the population as being entirely in accordance with the views of the 'Creator' – i.e. God, whom he viewed as being a 'benevolent' entity.

Malthus's influence on Darwin

Malthus's *Essay* had a profound influence on Darwin and would provide key building blocks when it came to the formulation of his great theory. Having read Malthus's *Essay on the Principle of Population*, Darwin declared that

> being well prepared to appreciate the struggle for existence which everywhere goes on from long-continued observation of the habits of animals and plants, it at once struck me that under these circumstances favourable variations [the occurrence of an organism in more than one distinct color or form[17]] would tend to be preserved, and unfavourable ones to be destroyed.[18]

Whereas Malthus has elegantly described the phenomenon of 'natural selection' (defined as the process whereby organisms better adapted to their environment tend to survive and produce more offspring,[19] though it was Darwin who was the first to coin this phrase), it was Darwin who went one step further, by asserting that 'the result of this would be the formation of new species'.[20] Said he, 'Here, then, I had at last got a theory by which to work …'.[21]

Darwin was aware that, without variation, there could be no divergence. However, both the mechanism by which variations occurred, and the mechanism by which species in which favourable variations had been preserved continued to 'diverge' in character, remained a mystery.[22] However,

> that they have diverged greatly is obvious from the manner in which species of all kind can be classed under a genera, genera under families, families under sub-orders, and so forth; and I can remember the very spot in the road, whilst in my carriage, when to my joy the solution occurred to me; and this was long after I had come to Down [Down House, his future marital home]. The solution, as I believe, is that the modified offspring of all dominant and increasing forms tend to become adapted to many and highly diversified places in the economy of nature.[23]

Here, Darwin is evidently postulating, as a complete or partial explanation for variation, that 'species of all kind' have some innate (but unexplained) ability to adapt to their environment. He would return to this matter in due course.

Chapter 12

Romance: Marriage: Darwin's Theory Takes Shape

In his letter to Henslow of 3 November 1838, Darwin indicates that he is on the trail of something quite remarkable.

> You may recollect how often I have talked over the marvellous fact of the species of birds being different, in those different islands of the Galapagos. Lately I have gained some curious facts, bearing on the same points, regarding the lizards & tortoises of those same islands, & now I want to know whether you can tell me anything about the plants. [In particular] … whether in casting your eye over my plants [i.e. those specimens which Darwin had sent home from the Galapagos], how many cases there are of *near* species, of the same genus; one species coming from one island & the other from a second island.[1]

On 12 November, Darwin wrote from Shrewsbury to Lyell to announce that he was shortly to be married.

> The lady is my cousin, Miss Emma Wedgwood, the sister of Hensleigh Wedgwood [born 1803], and of the elder brother [Josiah Wedgwood (II), born 1795, of Leith Hill Place, Surrey] who married my sister [Caroline, on 1 August 1837], so we are connected by manifold ties, besides on my part by the most sincere love and hearty gratitude to her for accepting such a one as myself.[2]

For Darwin, it seems, mutual love did come into the marital equation, after all! As for Emma, she said of Darwin

> He is the most open, transparent man I ever saw, and every word expresses his real thoughts. He is particularly affectionate and very nice to his father and sisters, and perfectly sweet tempered, and possesses some minor qualities that

add particularly to one's happiness, such as not being fastidious, and being humane to animals. We shall live in London, where he is fully occupied with being [Honorary] Secretary to the Geological Society and conducting a publication upon the animals of Australia [i.e. *The Zoology of HMS* Beagle *under the Command of Captain FitzRoy*].[3]

In that year of 1838, Darwin formulated some questions for a 'Mr Wynne' (unidentified) on the subject of inheritance, such as 'Are offspring like fathers or mothers?' and, 'When wild animals cross with tame does [their] offspring favour the former … [i.e. in their characteristics]?'[4]

Darwin reported to Emma on 2 January 1839, from the couple's intended new abode, which was to be 12 Upper Gower Street, Bloomsbury, London:

All my goods are in their proper places, and one of the front attics (henceforward to be called the Museum) is quite filled, but holds everything very well.[5]

In other words, it was here that Darwin would keep his specimens relating to natural history. On 20 January Darwin wrote to Emma to say

I was thinking this morning how it came that I, who am fond of talking and am scarcely ever out of spirits, should so entirely rest my notions of happiness on quietness and a good deal of solitude. But I believe the explanation is very simple. It is that during the five years of my voyage … the whole of my pleasure was derived from what passed in my mind while admiring views by myself, travelling across wild deserts or glorious forests, or pacing the deck of the poor little *Beagle* at night.[6]

Darwin and Emma were married at the Parish Church of St Peter in the Staffordshire village of Maer on 29 January.

On 5 February Emma wrote to her sister (Sarah) Elizabeth Wedgwood to say that her pianoforte had arrived at Gower Street and that she had 'given Charles a large dose of music every evening'.[7] Elizabeth, having recently visited Darwin and Emma at the Darwin family home at Shrewsbury, declared on 5 June, 'Charles [i.e. Darwin] goes to his own room to work after breakfast till 2 o'clock … '.[8]

That October, Darwin told Fox that Hensleigh Wedgwood had made

a curious discovery regarding our august family, which I must tell you, that a W. Darwin my great grandfather is described in the Phil. Transacts [*Philosophical Transactions*] for 1719, as a person of curiosity, who

> discovered the remains of a giant, evidently an Ichthyosaurus.- so that *we* have a right of hereditary descent to be naturalists & especially geologists.[9]

Here, Darwin has the wrong initial. It was, in fact, Robert Darwin (1682–1720), his great grandfather, who had discovered the skeleton in question.

Despite being a meticulous observer, Darwin was not always correct in his conclusions. He himself, confessed as much in respect of 'the parallel roads of Glen Roy' in the Lochaber region of the Scottish Highlands, which, from a distance, can easily be mistaken for man-made creations. Having put forward (in a paper published in *Philosophical Transactions*, 1839) the idea that the so-called 'roads' were caused by 'the action of the sea', he was obliged 'to give up this view when Swiss palaeontologist, glaciologist and geologist Louis Agassiz propounded his glacier-lake theory'. In essence, Agassiz proposed that, during some previous ice age, the glen had been dammed by a glacier and this, in turn, had created lakes which had varied in depth over time. At the perimeter of these ancient frozen lakes, the ice had sculpted out the underlying rock – hence the appearance of 'roads', which were, of course, not roads at all.[10]

When the Darwins' children arrived a governess was employed to look after them. On 27 December 1839, William Erasmus was born. The couple would subsequently have nine more children (three of whom would not survive to adulthood): Anne Elizabeth (2 March 1841–23 April 1851); Mary Eleanor (23 September 1842–6 October 1842); Henrietta Emma ('Etty', born 25 September 1843); George Howard (born 9 July 1845); Elizabeth (born 8 July 1847); Francis (born 16 August 1848); Leonard (born 15 January 1850); Horace (born 13 May 1851), and Charles Waring (6 December 1856–28 June 1858).

An illustration of how Darwin employed his enquiring mind and of his capacity for minute observation, is provided by an article which he wrote for the *Gardeners' Chronicle* (16 August 1841), a horticultural magazine founded that year. It covered every aspect of gardening and contained articles submitted by both gardeners and scientists. The article related to the manner in which 'humble-bees … bore holes in flowers and thus extract the nectar'.[11]

It was in 1842, for reasons of ill health, that Darwin decided to seek peace and quiet, and to this end he and Emma relocated, on 14 September, to Down House, situated near to Downe in Kent and only sixteen miles from the centre of the London. (The cause of Darwin's illness, which proved to be a chronic one, will be discussed shortly.) The village consisted of forty or so dwellings, the thirteenth-century Church of St Mary the Virgin, the George and Dragon Inn, and a butchers and bakers. A carrier journeyed to London once a week to collect special orders.[12]

Down House was a Georgian mansion, set in eighteen acres of land with a fifteen-acre meadow. Over the next four decades, the Darwins would enlarge it to include a new drawing room, billiard room, and extra bedrooms. The drawing room contained Emma's beloved Broadwood pianoforte, and a chaise longue, on which Darwin reclined as she played to him. Either that, or the couple would enjoy their daily, hotly-contested game of backgammon.[13] In the grounds, Darwin erected greenhouses, in order to cultivate and study tropical plants, such as orchids.

Down House became not only the family home, but also a nerve centre from which Darwin's tentacles reached out far and wide, to his natural-scientist friends (most notably, Lyell and Henslow) wherever in the world they might be. In this way he was able to draw on their knowledge of such diverse subjects as the temperatures of sea water at different depths on the west coast of South America; coral reefs; whirlwinds; earthquakes; the effects of glaciation; and the possible reasons for the different distribution of various species throughout the world. For the *Beagle* voyage had fuelled Darwin's interest in the natural world on a truly global scale. And just as fellow experts and enthusiasts were, almost invariably, willing to share their knowledge with him, so he was more than willing to reciprocate, as his immense, world-wide correspondence confirms.

Darwin wrote to Captain FitzRoy on 31 March 1843, happy that the latter had been appointed Governor of New Zealand, but sad that the captain would be 'leaving the country without me seeing you again'.[14]

Darwin's uncle and father-in-law Josiah Wedgwood (II), died on 12 July of that year after a long illness.

Two weeks later Darwin wrote to naturalist George Robert Waterhouse on the subject of how species were to be classified and what part, if any, the laws of 'God the Creator' should play in this process, whatever these laws might be.

> Linnaeus confesses profound ignorance. Most authors say it [i.e. the attempt at classification] is an endeavour to discover the laws according to which the Creator has willed to produce organized beings [those which are arranged into a structured whole[15]]; But what empty high-sounding sentences these are – it does not mean order in time of creation [i.e. the chronological order in which such 'beings' were 'created'], nor propinquity to any one type, as man. In fact it means just nothing. According to my opinion … classification consists in grouping beings according to their actual *relationship*, i.e. their consanguinity, or descent from common stocks.[16]

In other words Darwin completely rejected the notion of 'Creationism'.

In January 1844 Darwin told Hooker, who was then engaged in classifying and cataloguing the specimens of Galapagos flora which Darwin had sent him:

> Besides a general interest about the Southern lands, I have been now ever since my return engaged in a very presumptuous work & which I know no one

> individual who w[d] not say a very foolish one. I was so struck with distribution of Galapagos organisms, &c &c & with the character of the American fossil mammifers [mammals], &c &c that I determined to collect blindly every sort of fact, which c[d] bear any way on what are species .... At last gleams of light have come, & I am almost convinced (quite contrary to [the] opinion I started with) that species are not (it is like confessing a murder) immutable. Heaven forfend [avert] me from Lamarck['s] nonsense of a 'tendency to progression' 'adaptations from the slow willing of animals' &c, but the conclusions I am led to are not widely different from his – though the means of change are wholly so – I think have found out (here's presumption) the simple way by which species become exquisitely adapted to various ends.[17]

Jean Baptiste P. A. de Monet de Lamarck, to whom Darwin refers, was a French naturalist who was appointed Professor of Botany at the Jardin de Plantes, and subsequently Professor of Zoology at the Muséum National d'Histoire Naturelle in Paris.

Lamarck believed that simple organisms are continually created by 'spontaneous generation'; that species do not become extinct but, instead, change into other species; that species change their behaviour in response to environmental changes, and that this changed behaviour modifies their organs - the 'improved' structure being inherited by their offspring. To give an example, the giraffe has a long neck and long legs. By virtue of its great height, it is therefore able to reach up and graze off the leaves of high trees. According to Lamarck (about whom more will be said later), the giraffe has brought about these variations in its anatomy by its own volition - a proposition about which Darwin was highly sceptical.

Just as Darwin had hoped, a small army of experts was now at work on the specimens which he had collected from South America. For example, his specimens of Galapagos beetles (*Coleoptera* and *Heteromera*) were the province of naturalist George R. Waterhouse. As for Hooker, he and Darwin were more than professional colleagues. They had become friends who recommended reading matter to one another, and lent one another books on natural history

Darwin's hyperactive mind, in its unceasing quest for knowledge and explanation, roamed across the whole spectrum of natural history and attempted to answer such questions as why species were abundant in certain areas rather than in others? What factors limit the increase of a given species? What was the role of water or air currents in transporting species from one part of the planet to another? Why do sea shells found at opposite ends of the Panama Canal differ from each other? What causes earthquakes? Was the ability of a dog to identify a scent left by another animal an inherited quality?

Darwin wrote to zoologist and botanist the Reverend Leonard Jenyns, on 12 October to say,

> I have continued steadily reading & collecting facts on variation of domestic animals & plants & on the question of what are species; I have a grand body of facts & I think I can draw some sound conclusions. The general conclusion at which I have slowly been driven from a directly opposite conviction is that species are mutable [liable to change[18]] & that allied species are co-descendants of common stocks.[19]

In his letter to Hooker of 10 September 1845 Darwin states that comparative anatomist Professor Richard Owen

> is vehemently opposed to any mutability in species … . Lamarck is the only exception, that I can think of, of an accurate describer of species at least in the Invertebrate kingdom, who has disbelieved in permanent species, but he in his absurd though clever work has done the subject harm … .[20]

In early October, Darwin commiserated with Hooker, who had learnt that he had failed in his attempt to be elected to Edinburgh University's Chair of Botany.

When, in that same month, Darwin visited Chatsworth House in Derbyshire – seat of the Dukes of Devonshire – he was, in his words, 'like a child, transported with delight'. This delight was not, as might be guessed, with the house itself, but with 'the great Hot house [heated glasshouse, where plants are reared], and exotic plants building, & especially the water part, [which] is more wonderfully like tropical nature, than I could have conceived possible'.[21]

Emma's mother, Elizabeth Wedgwood, died on 31 March 1846. In that year, the first major extension was made to Down House, to include a schoolroom for the children and new accommodation for the household staff. The estimated cost was £300, and Darwin hoped that the 'Shrewsbury conclave' – i.e. his family - would not condemn him for such 'extreme extravagance', but, as he told his sister Susan, 'It seemed so selfish to make the house so luxurious for ourselves and not comfortable for our servants…'.[22]

When Charles Lyell received a knighthood, Darwin wrote on 24 September 1848 to congratulate him.[23]

Dr Robert Darwin died on 13 November. Said Darwin to Hooker,

> no one who did not know him would believe that a man above eighty-three years old could have retained so tender and affectionate a disposition, with all his sagacity unclouded to the last. I was at the time unable to travel, which added to my misery.[24]

It is a measure of the incapacitating nature of Darwin's illness, that he had been unable to attend the funeral of his beloved father, which must indeed have been a great sadness to him.

To Hooker, on 13 June 1849, Darwin described how variation in species made classification difficult.

> When the same organ is RIGOROUSLY compared in many individuals, I always find some slight variability, and consequently that the diagnosis of species from minute differences is always dangerous. Systematic work would be easy were it not for this confounded variation, which, however, is pleasant to me as a speculatist, though odious to me as a systematist.[25]

Again in a letter to Hooker, of 25 September 1853, he returned to the same theme.

> After describing a set of forms as different species, tearing up my MS [manuscript], and making them one species, tearing that up and making them separate, and them making them one again …, I have gnashed my teeth, cursed species, and asked what sin I have committed to be so punished.[26]

The problem confronting Darwin was that, however much species are subdivided into subspecies, variations in these subspecies continue to be observed.

The voyage of the *Beagle*, and the visit to the Galapagos Islands in particular, had provided Darwin with the vital clues to the fact that species were not immutable.

> From September 1854 I devoted my whole time to arranging my huge pile of notes, to observing, and to experimenting in relation to the transmutation [the action of changing or the state of being changed into another form, or the transformation of one species into another[27]] of species. During the voyage of the *Beagle* I had been deeply impressed by discovering in the [Argentinian] Pampean formation [consisting of loess - deposits of silt, which cover the Pampas - the extensive, treeless plains of South America] great fossil animals covered with armour like that on the existing armadillos; secondly by the manner in which closely allied animals replace one another in proceeding southward over the Continent; and thirdly, by the South American character of most of the productions of the Galapagos Archipelago, and more especially by the manner in which they differ slightly on each island of the group; none of the islands appearing to be very ancient in a geological sense.
>
> It was evident that such facts as these, as well as many others, could only be explained on the supposition that species gradually become modified; and the subject haunted me.[28]

Surely, Darwin should have rejoiced to think that he might have solved the problem of variation in the Galapagos; so why did he use the word 'haunted'? Was it because he anticipated, with alarm and dread, the proverbial coals of fire which, were he to publish his findings and conclusions, would surely be heaped upon his head by outraged Christian Biblical 'Creationists', whose default position, come what may, was to be found in the Old Testament's Book of Genesis. This states that, in a period of six days, God made the Earth, including its 'grass, herb-yielding seed, fruit trees, fowls, cattle, creeping things, beasts', and 'fish', until finally, on the sixth day 'God created *man* in his own image … male and female he created them'.[29]

In that year, 1854, Darwin was awarded the Royal Society's Royal Medal for his work on Cirripedia (barnacles).

> I am hard at work on my notes, collating & comparing them, in order [in] some 2 or 3 years to write a book with all the facts & arguments, which I can collect, for and versus the immutability of species.[30]

Many more honours would accrue to Darwin in the years to come. However this did not lead him to become arrogant. He was content to live a simple life, loathed ostentatiousness, dressed sombrely, did not crave material possessions, and continued to treat all and sundry with courtesy and consideration.

Meanwhile, he spent his time cultivating and cataloguing, breeding and cross-breeding plants and pigeons; amassing data relating to the cross-breeding of domestic dogs; exchanging information, ideas, and specimens of flora and fauna, not only with Hooker, Henslow, and many others in Britain, but also with his contacts in places as far afield as India, Harvard University (Cambridge, Massachusetts, USA), and Norway. He also found time to compose articles for the *Gardeners' Chronicle*.

In June 1855, Hooker indicated to Darwin that the two were as one in regard to the subject of transmutation. Said he:

> it is very easy to talk of the creation of a species in the Lyellian view of creation but the *idea* is no more tangible than that of the Trinity & to be really firmly & implicitly believed is neither more or less than a superstition – a believing in what the human mind cannot grasp.

Charles Lyell believed that 'new species were being continually created to replace those that had become extinct because of changes in the environment'.[31]

> It is much easier to believe with you in transmutation, until you work back to the vital spark – a *vis creatrix* [life-giving force] or whatever you may call it … [32]

Hooker continued: 'Lyell, it seems, still required some convincing about Darwin's great theory.'

Chapter 13

A Rival Appears on the Scene: Darwin's Hand Is Forced

In September 1855 an article was published in the *Annals and Magazine of Natural History*. Its author was Alfred Russel Wallace and it was to have profound implications for Darwin, who did not become aware of it until the summer of 1857.

Wallace, born at Usk, in Monmouthshire, South Wales on 8 January 1823, was the son of Thomas Vere Wallace, an attorney-at-law and his wife Mary Anne, née Greenell. He was therefore almost fourteen years younger than Darwin.

It was originally intended that Wallace should become a land surveyor, but in 1837, whilst he was residing with his brother, William in Bedfordshire he became interested in the science of geology. In this, and in other respects, his life was to run in a manner uncannily parallel to that of Darwin.

> My brother, like most land-surveyors, was something of a geologist, and he showed me the fossil oysters of the genus Gryphaea and the Belemnites … and several other fossils which were abundant in the chalk and gravel around Barton [(Barton-in-the-Clay, near Luton].[1]

Here, said Wallace

> during my solitary rambles I first began to feel the influence of nature … . At that time I hardly realized that there was such a science as systematic botany, that every flower and every meanest and insignificant weed had been accurately described and classified, and that there was any kind of system and order in the endless variety of plants and animals which I knew existed.[2]

Having, at the age of twenty-one, found employment as a teacher of English in Leicester, Wallace visited that city's 'very good town library', where

> perhaps the most important book I read was Malthus's *Principles of*

> *Population* [a volume also familiar to Darwin, as already mentioned] which I greatly admired for its masterly summary of facts and logical induction to conclusions.[3]

It was at Leicester that Wallace met the entomologist Henry Walter Bates, whose speciality was 'beetle collecting', and who also had 'a good set of British butterflies'.[4] It was here, also, that Wallace was 'first introduced' to the subject of 'psychical research'.[5]

Having relocated to Neath in South Wales, where he was employed as a surveyor, Wallace read *Vestiges of the Natural History of Creation* (published in 1844) by Robert Chambers, and 'was much impressed by it'.[6]

Robert Chambers

Robert Chambers (1802–71) was a Scottish publisher and geologist. In his book *Vestiges of the Natural History of Creation* (first published anonymously), he gave a lucid account of how

> a comparatively small variety of species is found in the older rocks … [whereas] ascending to the next group of rocks, we find the traces of life become more abundant, the number of species extended, and important additions made in certain vestiges of fuci [seaweed] or sea-plants, and of fishes.[7]
>
> We thus early begin to find proofs of the general uniformity of organic life over the surface of the earth at the time when each particular system of rocks was formed.[8]

Finally, Chambers was able to say

> We have now completed our survey of the series of stratified rocks, and traced in their fossils the progress of organic creation down to a time which seems not long antecedent to the appearance of man.[9]

Variation

Like Darwin, Chambers struggled to explain how 'new varieties', both in the animal and plant kingdoms, were created. 'We are ignorant of the laws of variety-production; but we see it going on as a principle in nature … .'[10]

The origin of life

'Whether mankind is of one or many origins', said Chambers, was 'still an open question'.

A 'Designer/Creator'

> It has been one of the most agreeable tasks of modern science to trace the

> wonderfully exact adaptations of the organization [i.e. structure] of animals to the physical circumstances amidst which they are destined to live. From the mandibles of insects to the hand of man, all is seen to be in the most harmonious relation to the things of the outward world, thus clearly proving that DESIGN presided in the creation of the whole – design again implying a designer, another word for a CREATOR.
>
> It would be tiresome to present in this place even a selection of the proofs which have been adduced on this point. The Natural Theology of [William] Paley, and the Bridgewater Treatises [of William Buckland, geologist, palaeontologist, and Dean of Westminster], place the subject in so clear a light that the general postulate may be taken for granted.[11]
>
> The face of God is reflected in the organization of man, as a little pool reflects the glorious sun.[12]

Clearly, therefore, Chambers subscribed, not to the theory of natural selection, but to that of a creator/designer – i.e. God. He concludes

> My sincere desire in the composition of the book (*Vestiges*) was to give the true view of the history of nature, with as little disturbance as possible to existing beliefs, whether philosophical or religious.[13]

In other words, to Chambers's way of thinking, scientific truth must always be subordinated to religious dogma.

To return to Wallace, he also mentions having read Darwin's *Journal*, of which he said

> As the Journal of a scientific traveller, it is second only to Humboldt's 'Personal Narrative' [which was one of Darwin's favourite books] as a work of general interest, [it is] perhaps superior to it.[14]

It was on account of having read these works by Darwin and Humboldt, said Wallace, that 'I owe my determination to visit the tropics as a collector'.[15] The outcome was that, on 20 April 1848, Wallace set sail from Liverpool en route to South America. Prior to his departure, however, 'the great problem of the origin of species was already distinctly formulated in my mind'.

> I believed the conception of evolution [the process by which different kinds of living organisms are thought to have developed and diversified from earlier forms during the history of the earth[16]] through natural law so clearly formulated in the [Robert Chambers's] *Vestiges* to be, so far as it went, a true one; and … I firmly believed that a full and careful study of the facts of nature

would ultimately lead to a solution of the mystery.[17]

Wallace describes 'four years' [of] wanderings in the Amazon valley' – a length of time comparable with that spent away from home by Darwin during the long voyage of HMS *Beagle*.[18] Also, like Darwin, Wallace became ill in the tropics, stating that, 'my health … has suffered so much by a succession of fevers and dysentery …[19]'.

On 6 August 1852, Wallace suffered another misfortune when the ship *Helen*, which was conveying him back to England, caught fire. 'My collections … were in the hold, and were irretrievably lost.' This included

> all my private collection of insects and birds … [which] comprised hundreds of new and beautiful species, which would have rendered (I had fondly hoped) my [display] cabinet, as far as regards American species, one of the finest in Europe.[20]

Fortunately, prior to the shipwreck, Wallace had managed to send a small part of his collection home to England. He had also managed to salvage some of his drawings of fish, and his diary. Said he, 'my collections had now made my name well known to the authorities of the Zoological and Entomological societies', a fact which gave him an automatic entrée into their scientific meetings.[21] Wallace finally arrived back to London on 5 October 1852.

Encouraged by Sir Roderick Murchison, President of the Royal Geographical Society, Wallace embarked, in early 1854 aboard the brig *Frolic* on another expedition, this time to the Far East.

It was at Sarawak, a state of the Island of Borneo, where Wallace was located from November 1854 until February 1856, that, in his words,

> I wrote an article which formed my first contribution to the question of the origin of species. I sent it to *The Annals and Magazine of Natural History*, in which it appeared in the following September (1855). Its title was 'On the Law which has regulated the Introduction of New Species,' which law was briefly stated … as follows: '*Every species has come into existence coincident both in space and time with a pre-existing closely-allied species*.' This clearly pointed to some kind of evolution. It suggested the *when* and the *where* of its occurrence, and that it could only be through natural generation, as was also suggested in the *Vestiges*; but the *how* was still a secret only to be penetrated some years later.[22]

Darwin told Hooker on 9 May 1856 that 'I had [a] good talk with Lyell about my species work, & he urges me strongly to publish something'.[23] Lyell clearly believed that, if Darwin prevaricated, then there was a real danger that he would be pre-empted and someone else would claim the credit for his great discovery of evolution by natural selection. Darwin took Lyell's advice, to 'write out my views [on the subject of evolution] pretty fully, and I began at once to do so …'.[24]

Wallace wrote to Darwin on 10 October, and expressed surprise at the lack of interest in his paper.[25] To this, Darwin replied,

> You say that you have been somewhat surprised at no notice having been taken of your paper [referred to above] in the Annals: I cannot say that I am; for so very few naturalists care for anything beyond the mere description of species.[26]

Darwin wrote graciously to Wallace on 1 May 1857 to say 'I agree to the truth of almost every word of your paper …',[27] whereupon the latter was 'much gratified'.[28] As Darwin states in his autobiography, this was not the last that he was to hear of Wallace.

> Early in the summer of 1858 Mr Wallace, who was then in the Malay archipelago, sent me an essay *On the Tendency of Varieties to Depart Indefinitely from the Original Type*; and this essay contained exactly the same theory as mine. Mr Wallace expressed the wish that if I thought well of his essay, I should send it to Lyell for perusal.[29]

On 18 June 1858 Darwin did as he was bid and forwarded Wallace's 'essay' to Lyell. He also admitted to Lyell that his (Lyell's) earlier warnings to him about the possible appearance of a rival had been fully justified.

> Some year or so ago, you recommended me to read a paper by [A. R.] Wallace in the Annals … . He has to day sent me the enclosed [i.e. the 'essay' referred to above] & asked me to forward it to you. It seems to me well worth reading. Your words have come true with a vengeance that I sh^{d} be forestalled … . I never saw a more striking coincidence, if Wallace had [been in possession of] my MS sketch written out in 1842 he could have not have made a better short abstract![30]

How Wallace arrived at the same conclusions as Darwin

Referring to the great question of the origin of present-day species, Wallace declared:

> My (1st) paper written at Sarawak rendered it certain to my mind that the change had taken place by natural succession and descent – one species becoming changed either slowly or rapidly into another. But the exact process of the change and the causes which led to it were absolutely unknown and appeared to be almost inconceivable.[31]

Wallace described how, when he wrote this paper, he was

> suffering from a sharp attack of intermittent fever, and every day during the cold and succeeding hot fits had to lie down for several hours, during which time I had nothing to do but to think over any subjects then particularly interesting me. One day something brought to my recollection Malthus's *Principles of Population*, which I had read about twelve years before. I thought of his clear exposition of 'the positive checks to increase [i.e. of population]' — disease, accidents, war, and famine — which keep down the population of savage races to so much lower an average than that of more civilized peoples. It then occurred to me that these causes [i.e. 'checks'] or their equivalents are continually acting in the case of animals also …[32]

Malthus, it will be remembered, was the very same person who had inspired Darwin. Wallace then posed the crucial question, 'Why do some die and some live? And the answer was clearly, that on the whole the best fitted live' or, in other words,

> the most healthy … the strongest, the swiftest … the most cunning … the best hunters or those with the best digestion; and so on. Then it suddenly flashed upon me that this self-acting process would necessarily *improve the race*, because in every generation the inferior would inevitably be killed off and the superior would remain — that is, *the fittest would survive*.[33]

Having, in this *eureka moment*, 'solved the problem of the origin of species', said Wallace,

> For the next hour I thought over the deficiencies in the theories of Lamarck and of the author of the *Vestiges*, and I saw that my new theory supplemented these views and obviated every important difficulty. I waited anxiously for the termination of my fit [of fever] so that I might at once make notes for a paper on the subject. The same evening I did this pretty fully, and on the two succeeding evenings wrote it out carefully in order to send it to Darwin by the next post …[34]

Darwin told Lyell on 25 June that he feared that Wallace might accuse him [Darwin] of plagiarism. Said he, 'I could send Wallace a copy of my letter to Asa Gray to show him that I had not stolen his doctrine.'[35] This was a reference by Darwin to his letter to Asa Gray of 20 July 1857, previously referred to, and in which was included an 'Enclosure', in which he explained his views on the subject of Natural Selection in great detail.

The outcome was that a gentlemen's agreement was reached, whereby Hooker and Lyell would forward papers containing 'the results of the investigations of two indefatigable naturalists, Mr Charles Darwin and Mr Alfred Wallace', to the Linnaean Society, This they did on 30 June 1858, together with a letter, which read as follows:

> These gentlemen having, independently and unknown to one another, conceived the same very ingenious theory to account for the appearance and perpetuation of varieties and of specific forms on our planet, may both fairly claim the merit of being original thinkers in this important line of inquiry; but neither of them having published his views, though Mr Darwin has for many years past been urged by us to do so, and both authors having now unreservedly placed their papers in our hands, we think it would best promote the interests of science that a selection from them should be laid before the Linnaean Society.

The 'papers' included Wallace's essay entitled 'On the Tendency of Varieties to Depart Indefinitely from the Original Type', and

> Extracts from a MS [Manuscript] work on Species, by Mr Darwin, which was sketched in 1839, and copied in 1844, when the copy was read by Dr Hooker, and its contents afterwards communicated to Sir Charles Lyell [together with] an abstract of a private letter addressed to Professor Asa Gray, of Boston, U.S., in October 1857, by Mr Darwin, in which he repeats his views, and which shows that these remained unaltered from 1839 to 1857.[36]

In other words Darwin had first formulated his theory no less than nineteen years previously!

Finally, at a meeting of the Linnaean Society held on 1 July 1858, Darwin's and Wallace's communications were read out (in their absence).[37]

To Hooker, on 13 July 1858, Darwin, as ever modest and self-effacing to a fault, wrote:

> I always thought it very possible that I might be forestalled, but I fancied that I had [a] grand enough soul not to care; but I found myself mistaken & punished; I had, however, quite resigned myself & had written half a letter to Wallace to give up all priority to him & sh[d] certainly not have changed [my mind] had it not been for Lyell's & yours quite extraordinary kindness.[38]

Finally, the following month, Wallace's 'Essay', together with an 'Abstract' from

Darwin's manuscript, were published in the *Journal of the Proceedings of the Linnean Society*.[39] However, said a disappointed Darwin, 'our joint productions excited very little attention … '.[40] Nevertheless, 'In September 1858, I set to work by the strong advice of Lyell and Hooker to prepare [for publication] a volume on the transmutation of species … .'[41]

Darwin told Wallace in April 1859:

> You are right, that I came to the conclusion that Selection was the principle of change from [a] study of domesticated productions; & then reading Malthus I saw at once how to apply the principle. Geographical Distrib. & Geological relations of extinct to recent inhabitants of S. America first led me to [the] subject. Especially [in the] case of Galapagos Isl[ds].
>
> I forget whether I told you that Hooker, who is our best British Botanist & perhaps best in [the] World, is a *full* convert, & is now going immediately to publish his confession of Faith [i.e. in Darwin's theory of natural selection] … . Huxley is changed & believes in mutation [i.e. variation] of species: [but] whether a *convert* to us [in respect of natural selection] I do not quite know. We shall live to see all the *younger* men converts.

This was a reference to biologist Thomas Henry Huxley, Professor of Natural History at the Royal School of Mines, London. Finally, Darwin paid Wallace this compliment: 'There have been few such noble labourers in the cause of Natural Science as you are.'[42]

Chapter 14

Labor Omnia Vincit

Meanwhile, Darwin's thoughts turned to how the sea could act as a vehicle for the transportation of seeds from one land mass to another. In November 1855, having experimented with various types of seeds to see how long they would float in salt water, he told Henslow

> I assume that half-dried plants with their fruit or pods would certainly float for several weeks, but having tried some 30 or 40 plants I have found only a single one which floated after a month's immersion, & most sink after one week. So that I am almost foiled about sea-transportal. However I may mention, the Capsicum seed germinated excellently after 137 days immersion in salt-water & Celery pretty well after the same period.[1]

On 9 December, Darwin wrote to Edgar Leopold Layard of the Cape of Good Hope Civil Service, and founder and first curator of the South African Museum, to say

> I have during many years been collecting all the facts & reasoning which I could, in regard to the variation & origin of species, intending to give [i.e. present], as far as lies in my power, the many difficulties surrounding the subject on all sides. One chief line of investigation naturally is concerned with the amount of variation of all our domestic animals.
>
> For various reasons, I have determined to work on pigeons, poultry, ducks & rabbits … .[2]

Said Darwin to Henslow on 22 January 1856

> I saw in the *Times* the death of your mother, but at so venerable an age that life can hardly be to any worth much further prolongation. In one sense I never knew what this greatest of losses is, for I lost my mother in very early childhood.[3]

The outcome of Darwin's insatiable appetite for his subject of natural history/natural science, was that he made numerous contacts in the field. For example, among the people with whom he corresponded in the year 1856 (including, of course, Huxley and Hooker), were Asa Gray (Fischer Professor of Natural History at Harvard University) and Syms Covington (Darwin's servant and assistant for eight years, during and after the voyage of the *Beagle*, who had emigrated to Australia in 1839).[4] Among his many and varied interests were the cross-breeding of ducks; the geographical distribution of crustacea; variations in the skeletons of different types of pigeon; the natural crossing of varieties of plants, including cacti; the trees of New Zealand, and what it was that enabled a glacier to flow or slide. (A selected list of those with whom Darwin corresponded, and also a list of his interests appears in the Appendix to this volume.)

On 3 May Darwin told Laurence Edmondston, fellow naturalist and physician at Unst, the most northerly of the Shetland Isles, 'I have devoted my whole life to do what little I could for our favourite pursuit of Natural History … .'[5]

The following month Darwin wrote to Fox to say that if he, Darwin, were to take Lyell's advice and publish his theory in the form of a 'Preliminary Essay … my work will be horribly imperfect & with many mistakes, so that I groan & tremble when I think of it'.[6]

Also in June, Darwin wrote to Hooker in respect of entomologist and conchologist Thomas V. Wollaston and his newly published book *On the Variation of Species, with a Special Reference to the Insecta; Followed by an Inquiry into the Nature of Genera*.[7] Wollaston did not share Darwin's belief in transmutation, and referred to 'the absurdity of the transmutational hypothesis'. Instead, he affirmed that 'the entire living panorama' was 'the work of a Master's [i.e. God, the Creator's] hand[8]'. Such a notion was, to Darwin, like a red rag to a bull.

> I have been very deeply interested by Wollaston's book, though I differ *greatly* from many of his doctrines. Did you ever read anything so rich, considering how very far he goes, as his denunciations against those who go further, 'most mischievous' 'absurd', 'unsound'. Theology is at the bottom at some of this. I told him he was like [John] Calvin burning a heretic. It is a very valuable & clever book in my opinion. He has evidently read very little out of his own line: I urged him to read the New Zealand Essay [A reference to J. D. Hooker's *Flora Novae – Zelandia*, in which the author examines 'contemporary views about the variability and fixity of plant species'].[9]

Darwin told Lyell on 5 July, 'I am *delighted* that I may say (with absolute truth) that my essay is published at your suggestion … .'[10] However, this statement proved to be premature, for that November, he wrote again to Lyell to say

> I am working very steadily at my big Book; I have found it impossible to publish any preliminary essay or sketch, but am doing my work as complete [ly] as my present materials allow without waiting to perfect them. And this much acceleration I owe to you.[11]

Said Darwin to Samuel P. Woodward, assistant in the Department of Geology and Mineralogy, British Museum, on 18 July 1856

> I am growing as bad as the worst about species & hardly have a vestige of belief in the permanence of species left in me, & this confession will make you think very lightly of me; but I cannot help it, such has become my honest conviction though the difficulties & arguments against such heresy are certainly most weighty.[12]

If the battle between Christian doctrine on the one hand and scientific evidence on the other was to be fought, there was no doubt in the mind of Darwin that science would emerge victorious. This did not mean, however, that the going would be easy. It was one thing to think private thoughts, write private notes, and share them with a select circle of friends. It was quite another to 'take Arms against a Sea of trouble'[13] by publishing his theory, thus incurring the wrath of the Church, criticism from unconvinced fellow scientists, and closer to home, the opprobrium of Emma and her Wedgwood family. Such was the dilemma in which Darwin found himself.

In late July Darwin told Hooker, 'Nothing is so vexatious to me, as so constantly finding myself drawing different conclusions from better judges than myself, from the same facts'.[14]

Darwin told Emma's sister-in-law Frances E. E. Wedgwood (née Mackintosh, wife of Hensleigh) in August that, although 'I should have liked to have [tried] my hand at Reviewing [i.e. books and articles]' he had rejected the notion. 'I have so many years work in prospect in my present book on species & varieties that I am not willing to give up my time to any other occupation …'[15]

'I do so wish I could understand clearly why you do not believe in accidental means of dispersion of plants',[16] wrote Darwin to Hooker in early 1857, indicating that the two men did not always see eye to eye.

Darwin wrote to US geologist and zoologist James D. Dana in April 1857 to say:

> [Professor] Owen has lately published a new Classification of mammals, taken from [i.e. based on] the structure of [their] Brain [s]; so great an authority ought to be right; but I cannot help always having doubts on a classification founded on one [single] character, however important.[17]

In that month of April Darwin told Lyell, 'though as a general rule I am much opposed to the Forbesian continental extensions, I have no objection whatever to its being proved in some cases'.[18] This was a reference to Edward Forbes, Professor of Natural History at Edinburgh University, who attempted to explain the distribution of flora and fauna by arguing that land masses must have been recently connected with each other. For example, said Forbes

> Although I have made icebergs and ice-flows the chief agents in the transportation of an Arctic flora southwards, I cannot but think that so complete a transmission of that flora as we find in the Scottish mountains, was aided perhaps mainly by land to the north, now submerged.[19]
>
> My main position may be stated in the abstract, as follows:
>
> *The specific identity, to any extent, of the flora and fauna of one area with those of another, depends on both areas forming, or having formed, part of the same specific centre, or on their having derived their animal and vegetable population by transmission through migration, over continuous or closely contiguous land, aided, in the case of alpine floras, by transportation on floating masses of ice.*[20]

Forbes also believed that species were originated by acts of divine creation, in specific centres, from which they migrated to the limits set by their conditions of existence.

Darwin told Hooker on 3 June,

> My observations, though on so infinitely a small scale, on the struggle for existence, begin to make me see a little clearer how the fight goes on: out of 16 kinds of seeds sewn on my meadow, 15 have germinated, but now they are perishing at such a rate that I doubt whether more than one will flower.[21]

Later that month Darwin asked Shropshire naturalist Thomas C. Eyton, to search 'in Ireland or elsewhere for any cases of Horses or Ponies with transverse bars [stripes] on [their] legs like those of zebra, or on shoulder & along the back, as with the ass.'[22]

On 20 July in a letter to Asa Gray, Darwin included an 'Enclosure', in which the following words appeared

> I think it can be shown that there is such an unerring power at work, or *Natural Selection* (the title of my Book), which selects exclusively for the good of each organic being.[23]

On 29 November, having been asked by Asa Gray to do so, Darwin defined what *he* understood by the term 'Natural Selection'. It meant, he said,

> The tendency to the preservation (owing to the severe struggle for life to which all organic beings at some time or generation are exposed) of the slightest variation in any part, which is of the slightest use or favourable to the life of the individual which has thus varied; together with a tendency to its inheritance.
>
> Any variation, which was of no use whatever to the individual, would not be preserved by this process of 'natural selection'.[24]

When his son Erasmus was in East Anglia with his tutor William Greive Wilson in February 1858 Darwin, anxious to miss no opportunity for scientific research, wrote to him to say,

> As Norfolk is near Suffolk, look out for me, whether there are near you any Suffolk Punches or [other] large Cart-Horses of a *Chestnut* colour; if so, please observe whether they have a dark stripe or band down the spine to root of tail; also for mere chance, whether any trace of a cross stripe on the shoulder, where the Donkey has [them], & any cross-stripes on the legs.[25]

Darwin's letter to Hooker of 23 June revealed, that on the domestic front, 'Poor dear Etty [his daughter, Henrietta] has been very seriously ill with Dipterithes [diphtheria] …'.[26] The health of the Darwin children will be discussed later.

To his son William, Darwin wrote on 22 September:

> If you go out shooting look at Birds' feet & see if any dirt sticks to them: I want to collect such dirt, & see, if by any splendid chance a plant would come up [derived from seed contained therein], for then could I not carry seeds across the sea![27]

Of an address, delivered by Professor Owen to the British Association for the Advancement of Science, of which he was president, held at Leeds in that same month, Darwin subsequently declared, 'He defines & further on amplifies his definition that Creation means a process he knows not what.'[28]

In October William commenced at Christ's College, Cambridge – his father's former college. The following February Darwin told Fox that William was 'very happy at Cambridge & he has changed into my old rooms …'.[29]

Writing from Down on 24 February 1859 Darwin told his son George that a billiard table had arrived on the premises: 'I heartily wish you & Willy were here to play with me. The whole affair has only cost £153. 18s. 0d.'[30] Clearly, the loving relationship that had existed between Darwin and his late father was now being mirrored by that which now existed between Darwin and his offspring.

On 20 October Darwin told Lyell, who persisted in favouring the 'Creationist' notion of evolution

> I have reflected a good deal on what you say on necessity of continued intervention of creative power. I cannot see this necessity; & its admission, I think would make the theory of [natural selection] valueless.[31]

Darwin was never afraid to challenge the theories of others, however eminent they might be. Yet, at the same time he was always ready to tell those, such as Lyell, how much he valued their friendship.

Another argument which the 'Darwinists' might have used against the 'Creationists' was put forward, a century and a half later, by US physician and evolutionary biologist Professor Randolph M. Nesse, who declared, 'No sensible person would have ever left the body the way it is. The human eye is a perfect example of why the body is not designed…', and Nesse drew attention to the fact a), that there is a 'blind spot' in the eye's field of vision and b), there is always the possibility of near or far-sightedness. He proceeds to cite further examples of 'poorly designed' bodily components – the appendix, wisdom teeth, and the female birth canal, which might almost have been intended to make birth difficult. The existence of these 'botched jobs', or 'built in vulnerabilities', can be explained only by natural selection.[32] Another argument, which Nesse might have put forward, was why would an intelligent 'god' trouble to create billions and billions of virtually identical organisms?

Darwin informed Professors Owen and Sedgwick, on 11 November, that he had instructed his publisher, John Murray, to send them each a copy of his forthcoming book *The Origin of Species*.[33]

In that year Darwin was awarded the Geological Society's Wollaston Medal 'for his geological work in South America, his theory of the origin and structure of coral reefs, and other contributions to geological science'.[34]

Finally, on 24 November, Darwin's magnum opus *The Origin of Species by Means of Natural Selection, or the Preservation of Favoured Races* [i.e. species] *in the Struggle for Life*, was published by John Murray.

Despite her misgivings about the theological implications of the work, Emma had previously helped her husband with the correction of the proofs.[35]

Chapter 15

The Origin of Species

Darwin's masterpiece was *The Origin of Species by Means of Natural Selection or the Preservation of Favoured Races in the Struggle for Life* commonly abbreviated to *Origin*. It was concerned not only with how animals and plants had evolved, but also with the mechanisms by which they had become distributed over the Earth's surface. Darwin's observations also included those of animals in their pre-natal state. He wrote:

> Hardly any point gave me any satisfaction when I was at work on the *Origin*, as the explanation of the wide difference in many classes between the embryo and the adult animal, and of the close resemblance of the embryos within the same class.[1]

In 1958, almost a century after it was published, Julian Huxley, grandson of Professor Thomas H. Huxley, posed the question

> Why is *The Origin of Species* such a great book? First of all, because it convincingly demonstrates the fact of evolution: it provides a vast and a well-chosen body of evidence showing that existing animals and plants cannot have been separately created in their present forms, but must have evolved from earlier forms by slow transformation. And secondly, because the theory of natural selection, which the *Origin* so fully and lucidly expounds, provides a mechanism by which such transformation could and would automatically be produced. Natural selection rendered evolution scientifically intelligible: it was this more than anything else which convinced professional biologists like Sir Joseph Hooker, T. H. Huxley, and Ernst Haeckel (German biologist and physician).[2]

In his 'Introduction' to *Origin*, Darwin states that, having returned from the *Beagle* voyage,

> It occurred to me, in 1837, that something might perhaps be made out on this question by patiently accumulating and reflecting on all sorts of facts which could possibly have any bearing on it. After five years' work I allowed myself to speculate on the subject, and drew up some short notes; these I enlarged in 1844 into a sketch of the conclusions, which then seemed to me probable: from that period to the present day I have steadily pursued the same object.

I have been urged to publish this Abstract … [i.e. *Origin*], as Mr Wallace, who is now studying the natural history of the Malay Archipelago, has arrived at almost the same general conclusions that I have on the origin of species.[3]

Chapter 1: Variation under Domestication
Of domestic pigeons, wrote Darwin,

> I have kept every breed which I could purchase or obtain … . The diversity of the breeds is something astonishing. Great as are the differences between the breeds of the pigeon, I am fully convinced … that they are all descended from the rock-pigeon [rock dove] (Colomba livia) … .[4]

The same applied to plants. For example

> the strawberry had always varied since it was cultivated, but the slightest varieties had been neglected. As soon, however, as gardeners picked out individual plants with slightly larger, earlier, or better fruit, and raised seedlings from them, and again picked out the best seedlings and bred from them, then (with some aid by crossing distinct species) those many admirable varieties of strawberry were raised which have appeared during the last half-century.[5]
>
> To sum up on the origin of our domestic races of animals and plants. Changed conditions of life are of the highest importance in causing variability, both by acting directly on the organisation, and indirectly by affecting the reproductive system. Variability is governed by many unknown laws … . Over all these causes of Change, the accumulative action of Selection, whether applied methodically and quickly, or unconsciously and slowly but more efficiently, seems to have been the predominant power.[6]

Here, Darwin postulates on the one hand, that variations are principally the result of changes in the environment, but on the other, admits that the question of variability remains largely an open one.

Chapter 2: Variation under Nature
Individual differences

> The many slight differences which appear in the offspring from the same parents … may be called individual differences.[7]

Summary

> We have … seen that it is the most flourishing or dominant species of the larger genera within each class which on an average yield the greatest number of varieties; and varieties, as we shall hereafter see, tend to become converted into new and distinct species. Thus the larger genera tend to become larger; and throughout nature the forms of life which are now dominant tend to become still more dominant by leaving many modified and dominant descendants. But by steps hereafter to be explained, the larger genera also tend to break up into smaller genera. And thus, the forms of life throughout the universe become sub-ordered into groups subordinate to groups.[8]

Chapter 3: Struggle for Existence
Darwin identifies climate, the availability of food, and the tendency of animals to prey upon one another as three of the factors which limit a species from increasing 'inordinately in numbers'. If, however, a species does multiply, despite these factors, then 'epidemics [of illness, causing disease and death] often ensue …'.[9]

Chapter 4: Natural Selection; or the Survival of the Fittest
Divergence of Character

> Mere chance … might cause one variety to differ in some character from its parents, and the offspring of this variety again to differ from its parent in the very same character and in a greater degree; but this alone would never account for so habitual and large a degree of difference as that between species.[10]

On the Degree to Which Organisation Tends to Advance

> Natural selection acts exclusively by the preservation and accumulation of variations, which are beneficial under the organic and inorganic conditions to which each creature is exposed at all periods of life. The ultimate result is that each creature tends to become more and more improved in relation to its

> conditions. This improvement leads to the gradual advancement of the organisation of the greater number of living beings throughout the world.[11]

Chapter 5: Laws of Variation

> I have hitherto sometimes spoken as if the variations … were due to chance. This, of course, is a wholly incorrect expression, but it serves to acknowledge plainly our ignorance of the cause of each particular variation.[12]

Although Darwin is anxious to explain why variation in species occurs, he is honest enough to admit that to do so was beyond the scope of present scientific knowledge.

Chapter 7: Miscellaneous Objections to the Theory of Natural Selection

> A serious objection has been urged by [German geologist and palaeontologist Heinrich G.] Bronn, and recently by [French physician and anthropologist Pierre Paul] Broca, namely, that many characters appear to be of no service whatever to their possessors, and therefore cannot have been influenced through natural selection.
>
> There is much force in the above objection. Nevertheless, we ought, in the first place, to be extremely cautious in pretending to decide what structures now are, or have formerly been, of use to each species.[13]
>
> Mr [St George J.] Mivart [biologist] is … inclined to believe that new species manifest themselves 'with suddenness and by modifications appearing at once'. This conclusion, which implies great breaks and discontinuity in the series, appears to me improbable in the highest degree.[14]

Chapter 8: Instinct

> No one will dispute that instincts are of the highest importance to each animal. Therefore there is no real difficulty, under changing conditions of life, in natural selection accumulating to any extent slight modifications of instinct which are in any way useful.[15]

Here, Darwin surmises that in the same way as physical characteristics may be inherited, so instinct may also be a heritable phenomenon.

Chapter 10: On the Imperfection of the Geological Record

> In the sixth chapter I enumerated the chief objections which might be justly urged against the views maintained in this volume. One, namely the distinctness

> of specific forms, and their not being blended together by innumerable transitional links, is a very obvious difficulty. [But] the number of intermediate varieties, which have formerly existed, [must] be truly enormous. Why then is not every geological formation and every stratum full of such intermediate links? The explanation lies, as I believe, in the extreme imperfection of the geological record.[16]
>
> Now let us turn to our richest geological museums, and what a paltry display we behold! That our collections are imperfect is admitted by every one. ... many fossil species are known and named from single and often broken specimens. Only a small portion of the surface of the earth has been geologically explored, and no part with sufficient care ... .[17]

It would not, therefore, have surprised Darwin to learn that more than a century and a half later 'intermediate (or 'missing') links' are still being discovered, virtually on a daily basis.

Chapter 11: On the Geological Succession of Organic Beings
On Extinction

> The extinction of species has been involved in the most gratuitous mystery. No one can have marvelled more than I have done at the extinction of species.[18]
>
> It is most difficult always to remember that the increase of every creature is constantly being checked by unperceived hostile agencies, and that these same unperceived agencies are amply sufficient to cause rarity, and finally extinction. So little is this subject understood, that I have heard surprise repeatedly expressed at such great monsters as the Mastodon and the ancient Dinosaurs having become extinct; as if mere bodily strength gave victory in the battle of life. Mere size, on the contrary, would in some cases determine, as has been remarked by Owen, quicker extermination from the greater amount of requisite food.[19]
>
> Thus, as it seems to me, the manner in which single species and whole groups become extinct accords well with the theory of natural selection. We need not marvel at extinction; if we must marvel, let it be at our own presumption in imagining for a moment that we understand the many complex contingencies on which the existence of each species depends.[20]

The question of why dinosaurs, the most spectacular creatures ever to have walked the Earth, became extinct will be discussed shortly.

Chapter 12: Geographical Distribution
The question arose, wrote Darwin, as to

> whether species had been created at one or more points of the earth's surface. Undoubtedly there are many cases of extreme difficulty in understanding how the same species could possibly have migrated from one point to the several distant and isolated points, where [that species is] now found.[21]

However, he pointed out that there were many factors which influenced the means by which living creatures could be dispersed.

Means of Dispersal

> Changes of climate must have had a powerful influence on migration. Changes of level in the land must have been highly influential. Where the sea now extends, land may at a former period have connected islands or possibly even continents together, and thus have allowed terrestrial productions to pass from one to the other.[22]

Darwin's own experiments indicated that seeds of certain plants might float on water for in excess of twenty-eight days, and yet still retain their power of germination. Seeds could also be spread by driftwood, via the carcasses of dead birds which floated on the sea with seeds contained in their crops; by living birds which may pass seeds through their intestines; or by seeds contained in soil which adheres to their claws and beaks.[23] In the Arctic and Antarctic regions seeds could be carried by icebergs which 'are sometimes loaded with earth and … brushwood …'.[24] Alternatively, 'Locusts are sometimes blown to great distances from the land; I myself caught one 370 miles from the coast of Africa.'[25]

Chapter 14: Mutual Affinities of Organic Beings …

> In this chapter I have attempted to show, that the arrangement of organic beings throughout all time in groups under groups – all naturally follow if we admit the common parentage of allied forms, together with their modification through variation and natural selection.[26]

Chapter 15: Recapitulation and Conclusion

> I have now recapitulated the facts and considerations which have thoroughly convinced me that species have been modified, during a long course of descent.

> This has been effected chiefly through the natural selection of numerous successive, slight, favourable variations …

This, then, was the essence of Darwin's great theory, which was based on the results of two decades of meticulous and painstaking research, fieldcraft, and experimentation. So far, so good. But Darwin went on to say that, in his opinion, the 'modification' (meaning, in this context, the transformation of an organism from its original anatomical form) of species, had also been

> aided in an important manner by the inherited effects of the use and disuse of parts; and in an unimportant manner, that is in relation to adaptive structures, whether past or present, by the direct action of external conditions and by variations which seem to us in our ignorance to arise spontaneously.[27]

Is it possible that 'modifications' – i.e. variations – in an organism could be brought about in such ways, and also that such variations could be passed down from one generation to another? This will shortly be discussed in more detail.

In an attempt (which proved unsuccessful, as will be seen) to pre-empt criticism and opprobrium from the Anglican Church, Darwin declared,

> I see no good reason why the views given in this volume should shock the religious feelings of any one. It is satisfactory, as showing how transient such impressions are, to remember that the greatest discovery ever made by man, namely, the law of the attraction of gravity, was also attacked by [German philosopher and mathematician Gottfried W.] Leibnitz 'as subversive of natural, and inferentially, of revealed religion'.[28]
>
> [However] … the chief cause of our natural unwillingness to admit that one species has given birth to clear and distinct species, is that we are always slow in admitting great changes of which we do not see the steps. The mind cannot possibly grasp the full meaning of the term of even a million years; it cannot add up and perceive the full effects of many slight variations, accumulated during an almost infinite number of generations.[29]
>
> Analogy would lead me one step farther, namely to the belief that all animals and plants are descended from some one prototype.[30]

This is the view commonly held today: that life on Earth first appeared about 3.8 billion years ago in the form of a single-celled organism.

Darwin completed *Origin* on a note of optimism

> as natural selection works solely by and for the good of each being, all corporeal and mental endowments will tend to progress towards perfection.[31]

In other words, as far as evolution was concerned, Darwin saw it as being applicable to the mind, as well as the body.

> Thus, from the war of nature, from famine and death, the most exalted object which we are capable of conceiving, namely, the production of the higher animals, directly follows. There is grandeur in this view of life, with its several powers, having been originally breathed by the Creator into a few forms or into one; and that, whilst this planet has gone cycling on according to the fixed law of gravity, from so simple a beginning endless forms most beautiful and most wonderful have been and are being evolved.[32]

So, surprisingly to Darwin's way of thinking, his great theory of evolution did not preclude the existence of a 'Creator'.

In the autumn of 1864, almost five years after the publication of *Origin*, Darwin described to Ernst Haeckel, Professor Extraordinarius of Zoology, University of Jena, how he had arrived at his theory.

> As you seem interested about the origin of the *Origin* & I believe do not say so out of mere compliment, I will mention a few points. When I joined the *Beagle* as Naturalist I knew extremely little about Natural History, but I worked hard. In South America three classes of facts were brought strongly before my mind: I^{stly} the manner in which closely allied species replace species in going Southward [i.e. from Brazil towards Tierra del Fuego]. 2^{ndly} the close affinity of the species inhabiting the Islands near to S. America to those proper to the Continent. This struck me profoundly, especially the difference of the species in the adjoining islets in the Galapagos Archipelago. 3^{rdly} the relation of the living Edentata [mammals such as sloths, anteaters and armadillos] & Rodentia [rodents] to the extinct species. I shall never forget my astonishment when I dug out a gigantic piece of armour like that of the living Armadillo.
>
> Reflecting on these facts & collecting analogous ones, it seemed to me probable that allied species were descended from a common parent. But for some years I could not conceive how each form became so excellently adapted to its habits of life. I then began systematically to study domestic productions [i.e. species which had been deliberately bred by man], & after a time saw clearly that man's selective power was the most important agent. I was prepared from having studied the habits of animals to appreciate the struggle for existence, & my work in Geology gave me some idea of the lapse of past time [i.e. the enormous amount of time which had elapsed since life on Earth began]. Therefore, when I happened to read 'Malthus on population' the idea of Natural

selection flashed on me. Of all the minor points, the last which I appreciated was the importance & cause of the principle of Divergence.[33]

Darwin defined 'Divergence' as 'the tendency in organic beings descended from the same stock to diverge in character as they become modified'.[34]

Finally, Darwin summarized his views thus:

> Although much remains obscure, and will long remain obscure, I can entertain no doubt, after the most deliberate study and dispassionate judgement of which I am capable, that the view which most naturalists until recently entertained, and which I formerly entertained – namely, that each species has been independently created – is erroneous. I am fully convinced the species are not immutable; but that those belonging to what are called the same genera are lineal descendants of some other and generally extinct species, in the same manner as the acknowledged varieties of any one species are the descendants of that species. Furthermore, I am convinced that Natural Selection has been the most important, but not the exclusive, means of modification.[35]

* * *

How was *Origin* received? Darwin told French zoologist and anthropologist J. L. Armand de Quatrefages de Bréau on 5 December 1859 that Lyell, Hooker, Huxley, and physician and naturalist William Benjamin Carpenter, amongst others, had all been converted to his way of thinking in respect of the 'mutability of species'.[36] Four days later Darwin's cousin, Francis Galton, wrote to congratulate him 'on the completion of your wonderful volume …'.[37] And, on 21 December, Darwin told Asa Gray that

> the 1st Edit [Edition] of 1250 copies was sold on [the] first day, & now my publisher is printing off as *rapidly as possible* 3000 more copies.[38]

Meanwhile, *Origin* had its detractors. For example, on the very day of its publication, Professor Sedgwick wrote disparagingly to Darwin to say

> I have read your book with more pain than pleasure. Parts of it I admired greatly; parts I laughed at till my sides were almost sore; other parts I read with absolute sorrow; because I think them utterly false & grossly mischievous … .[39]

To which, on 26 November 1859, Darwin responded, 'I cannot think a false theory would explain so many classes of facts, as the theory seems to me to do. But magna est veritas & thank God, prevalebit.'[40] Sedgwick was an evangelical Christian but

Darwin, in his letter, had demonstrated that he, too, had a good working knowledge of holy scripture: *Magna est veritas, et prevalebit* ('great is truth, and it prevails') being a quotation from the 'Vulgate', the commonly used Latin translation of the *Holy Bible*.[41]

Darwin's faith in his theory was unshakable. What the world now waited for with bated breath was for him to make a pronouncement on whether his theory applied to human beings and, if so, to what extent. The fact that Darwin believed that it did indeed apply to humans, and that he had come to this conclusion well before the publication of *Origin*, was revealed by him in his autobiography.

> As soon as I had become, in the year 1837 or 1838, convinced that species were mutable productions, I could not avoid the belief that man must come under the same law.[42]

He also declared that 'light would be thrown on the origin of man and his history' as a result of this book,[43] which was 'no doubt the chief work of my life'.[44]

For a full exposition by Darwin of his view on evolution in respect of man, however, the world would be obliged to wait more than a decade, until the publication of *The Descent of Man* in 1871.

Chapter 16

The Great Oxford Debate

With the onset of the new year work continued apace. On the sixth day of 1860 Darwin enquired of Thomas Bridges, a missionary in the Falkland Islands, as to the body language and facial expressions of the Fuegians and Patagonians. For example, do they

> nod their heads vertically to express assent, and shake their heads horizontally to express dissent? Do they blush … ? Do they sneer … ? Do they frown … ? Do they ever shrug their shoulders to show that they are incapable of doing or understanding anything?[1]

Darwin, in a letter of 18 January to Baden Powell, writer on theological topics and Savilian Professor of Geometry at Oxford University, quoted Sir John Herschel, who described the introduction of a new species as 'a natural in contradistinction to a miraculous process'.[2]

In April, wrote Darwin to Lyell,

> I can see no reason whatever for believing in such interpositions [of the Deity] in the case of natural beings, in which strange & admirable peculiarities have been naturally selected for the creature's own benefit.[3]

To Henslow, on 8 May, Darwin declared,

> I can perfectly understand Sedgwick or any one saying that nat. selection does not explain large classes of facts; but that is very different from saying that I depart from [the] right principles of scientific investigation.[4]

On 14 May Darwin wrote again to Henslow to say:

> I must thank you from my heart for so generously defending me as far as you could against my powerful attackers. Nothing which persons say hurts me for long, for I have entire conviction that I have not been influenced by bad

> feelings in the conclusions at which I have arrived. Nor I have I published my conclusions without long deliberation & they were arrived at after far more study than the [public] will ever know of or believe in. I am certain to have erred in many points, but I do not believe so much as Sedgwick & Co. think.[5]

Here, Darwin was implying that although others may have been motivated by 'bad feelings' towards him, he entertained no such animosity towards them. This situation would change, however, particularly in respect of one individual, namely Richard Owen.

Four days later Darwin informed Wallace that he [Darwin] was under the proverbial cosh from his critics.

> The attacks have been heavy & incessant of late. Sedgwick & Prof. [Clark] attacked me savagely at Cambridge [Philosophical Society] But Henslow defended me well, though [he is] not a convert [to Darwin's theory].- Phillips has since attacked me in [a] Lecture at Cambridge. Sir W. Jardine in [*Edinburgh New Philosophical Journal*]. Wollaston in [*Annals and Magazine of Natural History*]. A. Murray before [the] Royal [Society] of Edinburgh – Haughton at [Geological Society] of Dublin – Dawson in Canadian [*Naturalist*] Magazine, and *many others*.[6]

The persons referred to above were William Clark, clergyman and Professor of Anatomy at Cambridge University; John Phillips, Professor of Geology at Oxford University; William Jardine, 7th Baronet, naturalist and founder in 1841 of the *Annals and Magazine of Natural History*; Thomas Vernon Wollaston, entomologist and conchologist; Andrew Murray, entomologist and botanist and Assistant Secretary to the Royal Horticultural Society; Samuel Haughton, clergyman and paleobotanist, Registrar of the Medical School, Dublin, and John William Dawson, Professor of Geology and Principal of McGill University, Montreal, Canada.

On 22 May Darwin told Asa Gray that

> the most serious omission in my book was not explaining how it is … that all forms do not necessarily advance, how there can now be *simple* organisms still existing [i.e. which have not evolved despite the passage of time].[7]

As regards the controversy occasioned by the publication of his book, *The Origin of Species*, Darwin declared sanguinely,

> if I had not stirred up the mud someone else would very soon; so that the sooner the battle is fought the sooner it will be settled, not that the subject will be settled in our lives' times. It will be an immense gain, if the question

becomes a fairly open one; so that each man may try his new facts on it pro & contra.[8]

* * *

It was in Oxford, in the summer of 1860, at a meeting of the British Association for the Advancement of Science (held from 26 June to 3 July 1860), that the protagonists in the drama met head on: the 'Darwinists', on the one hand, represented principally by Huxley and Hooker, and ranged against them, the 'Creationists', represented by Professor Richard Owen, the Reverend Samuel Wilberforce (Bishop of Oxford), and Robert FitzRoy (Darwin's former captain on the *Beagle*, who was now an admiral), on the other. Darwin himself was unable to attend due to ill health, and Wallace had not returned from the Far East.

The following account of the meeting 'has been drawn from the London literary magazine the *Athenaeum* (not to be confused with the London club of that name), which provided the most complete contemporary report of the meeting and which Darwin himself read'.[9] (In fact, the *Athenaeum* published two reports, one on 7 July and the other a week later.)

> *Athenaeum*, 7 July 1860. Zoology and botany, including physiology. President: John Stevens Henslow. 'On the Final Causes of Sexuality of Plants, with particular Reference to Mr Darwin's work "On the Origin of Species by Natural Selection" by Dr DAUBENY'.

At the meeting, Charles G. B. Daubeny, Professor of Botany at Oxford University, stated that

> Whilst … he gave his assent to the Darwinian hypothesis … he wished not to be considered as advocating it to the extent to which the author seems disposed to carry it.

Daubeny, in other words, preferred to 'sit on the fence'. Whereupon, Professor Huxley,

> having been called on by the Chairman, deprecated any discussion of the general question of the truth of Mr Darwin's theory. He felt that a general audience, in which sentiment would unduly interfere with intellect, was not the public before which such a discussion should be carried on.

Professor Owen,

> Whilst giving all praise to Mr Darwin for the courage with which he had put forth his theory, he felt it must be tested by facts. As a contribution to the facts

by which the theory must be tested, he would refer to the structure of the highest Quadrumana [primate, other than a human, having all four feet modified as hands] as compared with man. Taking the brain of the gorilla, it presented more differences, as compared with the brain of man, than it did when compared with the brains of the very lowest and most problematical form of the Quadrumana. The deficiencies in cerebral structure between the gorilla and man were immense. The posterior lobes of the cerebrum in man presented parts which were wholly absent in the gorilla. The same remarkable differences of structure were seen in other parts of the body ....

Professor Huxley

begged to be permitted to reply to Prof. Owen. He denied altogether that the difference between the brain of the gorilla and man was so great as represented by Prof. Owen, and appealed to the published dissections of [Professor Friedrich] Tiedemann [German anatomist and physiologist] and others. From the study of the structure of the brain of the Quadrumana, he maintained that the difference between man and the highest monkey was not so great as between the highest and the lowest monkey. He maintained also, with regard to the limbs, that there was more difference between the toeless monkeys and the gorilla than between the latter and man. He believed that the great feature which distinguished man from the monkey was the gift of speech.[10]

Athenaeum, 14 July 1860, Zoology and botany, including physiology. 'On the Intellectual Development of Europe, considered with Reference to the Views of Mr Darwin and others, that the Progression [evolution] of Organisms is determined by Law,' by Prof. DRAPER, MD, of New York.

In his address, Dr John W. Draper, Professor of Chemistry at the University of the City of New York and President of its Medical School, described the 'doctrine of the immutability of species' as 'fanciful'. On this vitally important point, he was therefore in agreement with Darwin.

The Bishop of Oxford, the Reverend Samuel Wilberforce,

stated that the Darwinian theory, when tried by the principles of inductive science, broke down. The facts brought forward did not warrant the theory. The permanence of specific forms was a fact confirmed by all observation. The remains of animals, plants, and man found in those earliest records of the human race — the Egyptian catacombs, all spoke of their identity with existing

> forms, and of the irresistible tendency of organized beings to assume an unalterable character. The line between man and the lower animals was distinct: there was no tendency on the part of the lower animals to become the self-conscious intelligent being, man; or in man to degenerate and lose the high characteristics of his mind and intelligence. He [the bishop] was glad to know that the greatest names in science were opposed to this theory, which he believed to be opposed to the interests of science and humanity.

Professor Huxley

> defended Mr Darwin's theory from the charge of its being merely an hypothesis. He said it was an explanation of phenomena in Natural History .... Darwin's theory was an explanation of facts; and his book was full of new facts, all bearing on his theory. Without asserting that every part of the theory had been confirmed, he maintained that it was the best explanation of the origin of species which had yet been offered.

Admiral Robert FitzRoy

> regretted the publication of Mr Darwin's book, and denied Prof. Huxley's statement that it was a logical arrangement of facts.

Dr Hooker,

> being called upon by the President to state his views of the botanical aspect of the question, observed that the Bishop of Oxford having asserted that all men of science were hostile to Mr Darwin's hypothesis, whereas he himself was favourable to it, he could not presume to address the audience as a scientific authority.

This was clearly sarcasm on Hooker's part, he being one of the greatest scientific experts of the age in his chosen field of botany.

> As, however, he had been asked for his opinion, he would briefly give it. In the first place, his Lordship [the bishop], in his eloquent address, had, as it appeared to him, completely misunderstood Mr Darwin's hypothesis: his Lordship intimated that this maintained the doctrine of the transmutation of existing species one into another, and had confounded this with that of the successive development of species by variation and natural selection. The first of these doctrines was so wholly opposed to the facts, reasonings and results of Mr

> Darwin's work, that he [Hooker] could not conceive how any one who had read it could make such a mistake, the whole book, indeed, being a protest against that doctrine.

In other words, Hooker was implying that Bishop Wilberforce had either not taken the trouble to read *Origin*; or had read it, but failed to interpret the facts contained within it correctly.

> Now … that Mr Darwin had published it, he had no hesitation in publicly adopting his hypothesis, as that which offers by far the most probable explanation of all the phenomena presented by the classification, distribution, structure, and development of plants in a state of nature and under cultivation; and he should, therefore continue to use his [Darwin's] hypothesis as the best weapon for future research, holding himself ready to lay it down should a better be forthcoming, or should the now abandoned doctrine of original creations regain all it had lost in his experience.[11]

Chapter 17

Aftermath of the Great Debate

In a letter to Asa Gray, dated 22 July 1860, Darwin paid this tribute to those who had supported him at the great Oxford debate.

> I see most clearly that my book would have been a dead failure, had it not been for all the generous labour bestowed on it (not for my sake, but for the subject sake) by yourself, Hooker, Huxley & Carpenter [William B. Carpenter, physician and naturalist]; & to these names I hope soon Lyell's may be added.[1]

However, repercussions of the debate continued to rankle with Darwin, even though he had not been present at it in person. This is evident in a letter which he wrote on 3 August 1860 to his publisher John Murray, in respect of a review which Bishop Samuel Wilberforce had written of *Origin*, which was published, anonymously, in the July 1860 issue of the *Quarterly Review.*

> The Bishop makes me say several things which I do not say, but these very clever men think they can write a review with a very slight knowledge of the Book reviewed or subject in question.[2]

On 23 April 1861 Darwin told Hooker

> In simple truth I am become quite demoniacal about Owen, worse than Huxley … . I shall never forget his [Owen's] cordial shake of the hand when he was writing as spitefully as he possibly could against me.[3]

This was a reference to a meeting between Darwin and Professor Owen soon after the publication of *Origin*. Darwin was not known for being antagonistic towards his detractors, but Owen's underhand attacks exhausted his patience and filled him with disgust and contempt.

To Hooker on 25/26 January 1862, Darwin once again expressed his frustration with Owen.

> By the way Huxley tells me that Owen goes in for progressive development in the 2d. Edit. of his Palaeontology, pooh-pooing natural selection. I am quite ashamed how demoniacal my feelings are towards Owen.[4]

This was a reference to *Palaeontology or a Systematic Summary of Extinct Animals and their Geological Relations*, published in 1860.

Darwin told Armand de Quatrefages on 11 July how, in respect of *Origin*,

> I have been atrociously abused by my religious countrymen; but as I live an independent life in the country, it does not in the least hurt me in any way. except indeed when the abuse comes from an old friend, like Prof. Owen … .[5]

On 4 April 1863 a paper entitled, 'Introduction to the Study of the Foraminifera [single-celled, planktonic marine animals]' by William Carpenter was published in the *Athenaeum*, and reviewed by an anonymous person whom Darwin identified as Professor Owen.[6] Not only that but, in his review, the reviewer took the opportunity to criticize Darwin's theory of evolution, which led the latter (in a letter to the *Athenaeum's* editor), to respond thus:

> Your reviewer thinks that the weakness of my theory is demonstrated because existing Foraminifera are identical with those which lived at a very remote epoch. So little do we know of the conditions of life all around us, that we cannot say why one native weed or insect swarms in numbers, and another closely allied weed or insect is rare. Is it then possible that we should understand why one group of beings has risen in the scale of life during the long lapse of time, and another group has remained stationary?[7]

Darwin elaborated upon this point further in his letter of 22 May to George Bentham, botanist and President of the Linnean Society of London.

> in judging the theory of natural selection, which implies that a form will remain unaltered unless some alteration be to its benefit, is it so very wonderful that some forms should change much slower & much less, & some few should have changed not at all under conditions which to us (who really know nothing [of] what are the important conditions [i.e. circumstances]) seem very different.[8]

On 19 June Darwin quoted Samuel Wilberforce, the Bishop of Oxford, as having said that 'he believed that the *Origin* was the most illogical book ever published'.[9]

* * *

Meanwhile, to economist and politician Henry Fawcett on 6 December 1860, Darwin had made this disclosure in regard to his *modus operandi.*

> As you seem so kindly interested in my work, I may mention that I believe that the key of my work was gained by an unusually inductive line of research ['induction' being defined as the inference of a general law from particular instances].[10]

From Ternate – an island in the Indonesian archipelago of the Moluccas – on Christmas Eve, Wallace wrote to Henry W. Bates (who had been his companion on the expedition to South America) in fulsome praise of Darwin.

> Mr Darwin has created a new science and a new philosophy; and I believe that never has such a complete illustration of a new branch of science been due to the labours and researches of a single man.[11]

Darwin advised Hooker on 4 February 1861 to take life more gently.

> Be idle; but I am a pretty man to preach, for I cannot be idle, much as I wish it & am never comfortable except when at work. The word Holiday is written in a dead language for me, & much I grieve at it.[12]

To Armand de Quatrefages, on 25 April, Darwin declared of *Origin*, 'My views spread slowly in England & America; and I am much surprised to find them most commonly accepted by Geologists, next by Botanists and least by Zoologists.'[13]

To his son William, on 9 May, Darwin wrote, 'I have not had one game of Billiards since the Boys [his other sons, George and Francis] were here; indeed the Table has been covered with skeletons of Cocks & Hens & has been very useful for that purpose.'[14]

When Henslow died on 16 May Darwin told Hooker, 'I fully believe a better man never walked this earth.'[15]

To Sir John Herschel on 23 May Darwin wrote:

> The point which you raise on intelligent design has perplexed me beyond measure … . One cannot look at this Universe with all living productions &

> man without believing that all has been intelligently designed; yet when I look to each individual organism, I can see no evidence of this.[16]

The following day Darwin told Bartholomew J. Sulivan, naval officer and hydrographer (and lieutenant on the famous voyage of HMS *Beagle*, 1831–36):

> FitzRoy was so kind as to send me the last *London Review* & I read the article on Genesis. I cannot say that it all satisfied me … . But I am weary of all these attempts to reconcile, what I believe to be irreconcilable.[17]

Darwin had long regarded as irreconcilable his theory of evolution by natural selection, on the one hand, and the Biblical account of Creation as described in the Book of Genesis, on the other.

To Frances Julia Wedgwood (daughter of Hensleigh and his wife Frances) on 11 July Darwin returned to the question of 'design', this time in respect of the entire universe.

> The mind refuses to look at this universe, being what it is, without having been [i.e. without believing it to have been] designed; yet, where one would most expect design, viz. in the structure of a sentient being, the more I think on the subject, the less I can see proof of design.[18]

And he told Asa Gray on 17 September:

> I have lately been corresponding with Lyell, who, I think, adopts your idea of the stream [i.e. continuous occurrence] of variation [of species] having been designed … . I must think that it is illogical to suppose that variations which Nat. Selection, preserves for the good of any being, have been designed.[19]

In October Darwin wrote again to Gray, this time in respect of the anatomical structure of the nose.

> I should believe it to have been designed (as I did formerly each part of each animal) until I saw a way of its being formed without design, & *at the same time* saw in its whole structure evidence of its having been produced in a quite distinct manner, i.e. by descent from another cream-jug whose nose subserved, perhaps, some quite distinct use.[20]

(Gray had previously joked about a monster with a 'cream-jug' nose.)

On 30 November Darwin received a letter from Wallace sent from Sumatra, congratulating him on *Origin*, and in particular for 'both the attractive manner in which you have treated the subject & the clearness with which you have stated & enforced the arguments …'.[21]

Of the work of others Darwin could be a stern critic. For example, in December, referring to *Études sur la géographie botanique de l'Europe, et en particulier sur la végetation du plateau central de la France* by Henri Lecoq, Professor of Science at the University of Clermont-Ferrand, France, published in nine volumes between 1854 and 1858, Darwin declared: 'Lecoq ['s] is a miserable book, dreadfully spun out, with maudlin speculations & a great dearth of precise facts … .'[22]

To Asa Gray on 22 January 1862 Darwin wrote in reference to the American Civil War (1861–65):

> I have begun to think whether it would not be well for the peace of the world, if you [the Northern States, or Union, and the Southern States, or Confederacy] were split up into two or three nations. On the other hand I cannot bear the thought of the slave-holders [who existed predominantly in the southern states] being triumphant … .[23]

To the Reverend Charles Kingsley on 6 February, Darwin wrote on the question of

> the genealogy of man [line of evolutionary development from earlier forms][24] to which you allude. It is not so awful & difficult to me, as it seems [to] most, partly from familiarity & partly, I think, from having seen a good many Barbarians. I declare the thought, when I first saw in T. [Tierra] del Fuego a naked painted, shivering hideous savage, that my ancestors must have been somewhat similar beings, was at that time as revolting to me, nay more revolting than my present belief that an incomparably more remote ancestor was a hairy beast. Monkeys have downright good hearts, at least sometimes, as I could show, if I had space. I have long attended to this subject, & have materials for a curious essay on Human expression, & a little on the relation in mind [i.e. of the mind] of man to the lower animals. How I sh[d] be abused if I were to publish such an essay!
>
> It is very true what you say about the higher races of men, when high enough, replacing & clearing off the lower races. In 500 years how the Anglo-Saxon race will have spread & exterminated whole nations; & in consequence how much the Human race, viewed as a unit, will have risen in rank.[25]

At first glance the contents of the above paragraph come as a shock. How could Darwin, this kindly and humane man who abhorred slavery and any kind of cruelty, speak about the extermination of whole nations? But it would be a mistake to assume from this that, just because Darwin foresaw such an eventuality, he necessarily condoned it, or that, as a scientist, he was doing anything other than predicting the future in the light of what had happened in the past.

When he wrote to Hooker on 18 March 1862 Darwin was still wrestling with the nagging, yet seemingly inexplicable, phenomenon of variability, which he had attempted on several occasions to explain, but without any true conviction. I think, said he, 'that all variability is due to changes in the condition of life … [which] affect in an especial manner the reproductive organs, those organs which are to produce a new being'.[26] Darwin had expressed this view previously, and was doubtless aware that it was pure speculation on his part.

In early April Wallace sent Darwin 'a wild honeycomb from the island of Timor, not quite perfect but the best I could get'.[27] That spring Wallace returned to England, having spent eight years in the Far East.[28]

To Asa Gray in June, Darwin wrote:

> I received 2 or 3 days ago a French Translation of the Origin by a Madelle Royer [Clémence Auguste Royer, French author and economist], who must be one of the cleverest & oddest women in Europe: is [an] ardent Deist & hates Christianity, & declares that natural selection & the struggle for life will explain all morality, nature of man, [politics] &c &c!!! She makes some very curious & good hits [i.e. points], & says she shall publish a book on these subjects, & a strange production it will be.[29]

Darwin is doubtless grateful to Mademoiselle Royer for her support, but too much of a gentleman to expose her more extravagant claims on his behalf to ridicule.

Emma demonstrated that she was totally in agreement with her husband when she wrote to T. G. Appleton on 28 June:

> We shall rejoice at the termination of the [American Civil] war & [even] if we cannot hope to see Slavery abolished I think it must at all events be prevented from Spreading.[30]

Darwin, in effervescent mood, told Hooker on Christmas Eve:

> And now I am going to tell you a most important piece of news!! I have almost resolved to build a small hot-house: my neighbours really first-rate gardener has suggested it & offered to make me plans & see that it is well done … it will be grand amusement for me to experiment with plants.[31]

Following the discovery in Southern Germany of the fossilized bird *Archaeopteryx* (about which more will be said shortly), Darwin declared, 'It shows how little we know [of] what lived during former times.'[32]

On 7 January 1863 Darwin told James D. Dana that he had been told by palaeontologist and botanist Hugh Falconer, that Owen had 'not done the work well' in respect of an examination which the latter had performed on the fossil of *Archaeopteryx*.[33] This prompted him (Darwin) to write to Falconer to enquire, 'Has God demented Owen, as a punishment for his crimes, that he should overlook such a point?'[34] (It was alleged that the 'jaw with teeth' which Falconer described to Darwin in a previous letter, did not actually belong to *Archaeopteryx* at all; a point which Owen appears to have overlooked.[35])

Less than three weeks later, on the 26th, Darwin congratulated entomologist Henry W. Bates on his marriage and declared, 'Judging from my own experience [of that institution] it is [i.e. provides] the best & almost only chance for what share of happiness this world affords.'[36]

Darwin displayed another of his talents on 7 February when he wrote in French, to French botanist Charles V. Naudin, thanking him for supplying information on the cross-breeding of various types of melon.[37]

Throughout his working life Darwin had likewise been generous in giving his time and sharing his knowledge with others, as his immense correspondence testifies. However, he was not always able to satisfy their curiosity. For example, in 1882 he told Scottish physician and surgeon Robert S. Skirving of Edinburgh, 'I am sorry to say that I know nothing of the habits of earwigs.'[38]

On 13 March Darwin declared to Hooker, in respect of Lyell, who evidently had remained a Creationist, 'I feel sure that at times he no more believed in Creation than you or I.'[39]

A week later, Darwin wrote to Asa Gray 'to thank you in my dear little man's name for two precious stamps'. This was a reference by Darwin to his twelve-year-old son, Leonard, who was a collector of postage stamps.[40]

On 2 July Darwin advised John Scott, foreman of the Propagating Department at the Royal Botanic Garden, Edinburgh in respect of the latter's botanical experiments, to stress the necessity for care, honesty and scrupulousness when it came to research.

> By no means modify even in the slightest degree any result. Accuracy is the soul of Natural History. It is hard to become accurate; he who modifies a hair's breadth will never be accurate. It is a golden rule, which I try to follow, to put every fact which is opposed to one's preconceived opinion in the strongest light. Absolute accuracy is the hardest merit to attain & the highest merit. Any deviation is ruin.[41]

Throughout the year 1864 Darwin was preoccupied mainly with the subject of botany, having an essay entitled 'The Movements and Habits of Climbing Plants' published in the *Journal of the Linnaean Society*. On 30 November, in his absence, he was awarded the Copley Medal by the Royal Society of London for his 'important Researches in Geology, Zoology, and Botanical Physiology'.[42]

To Lyell, on 22 January 1865, Darwin wrote, with reference to variations found in humming birds on the one hand and orchids on the other, 'The more I work the more I feel convinced that it is by the accumulation of such extremely slight variations that new species arise.'[43]

Darwin told Hooker on 9 February of the horror which he felt at what he presumed was the

> certainty, of the sun some day cooling & we all freezing. To think of the progress of millions of years, with every continent swarming with good & enlightened men all ending in this; & with probably no fresh start until this our own planetary system has been again converted into red-hot gas. Sic transit gloria mundi ['So passes away the glory of the world'], with a vengeance.[44]

Robert FitzRoy, Vice Admiral (retired), Fellow of the Royal Society and former commander of HMS *Beagle*, committed suicide on 30 April.

In late September Darwin told Hooker, following the death of the latter's father on 12 August, 'I do not think anyone could love a father much more than I did mine & I do not believe three or four days ever pass without my still thinking of him … .'[45]

Darwin wrote to Hooker on 22 December saying,

> I have been so careless I have lost several diplomas & now I want to know what [Societies] I belong to, as I observe every[one] tacks their titles to their names in the Catalogue of The Royal [Society].[46]

It is interesting to note how Darwin could be so scrupulously meticulous on the one hand, and yet so absent-minded on the other!

To 'a local landowner' the following year, 1866, an outraged Darwin wrote:

> As you are now so little on your Farm, you may not be aware that the necks of your horses are badly galled [chafed] … . I must for the sake of humanity attend to this. I sincerely hope that you will at once make enquiries & give strict orders to your Bailiff not to work any horse with a wounded neck.[47]

In his time, Darwin must have impaled hundreds of insects on tiny pins, shot and stuffed countless birds, bottled in preservative a myriad of small creatures, but this was all in the cause of science. The serious neglect of a horse, however, was something that this humane and sensitive man was not prepared to tolerate.

In reply to Hooker's statement that 'all botanists would agree that many tropical plants could not withstand a somewhat cooler climate', Darwin replied, on 28 February 1866, 'I have come not to care at all for general beliefs without the special facts.'[48]

On 28 April Emma described how her husband had embarked on a rare and special outing.

> Charles went last night to the Soirée at the Royal Society, where assemble all the scientific men in London. The President presented him to the Prince of Wales. There were only three presented, and he was the first.[49]
>
> That spring, Darwin was hard at work preparing for the production of the 4th edition of *Origin*.

To Wallace, on 5 July, Darwin wrote:

> I fully agree with all that you say on the advantages of H. Spencer's excellent expression of 'the survival of the fittest'. I wish I had received your letter two months ago for I would have worked in [i.e. included] 'the survival etc' often in the new edition of the *Origin* which is now almost printed off …. The term Natural selection has now been so largely used abroad & at home that I doubt whether it could be given up, & with all its faults I should be sorry to see the attempt made. Whether it will be rejected must now depend on the 'survival of the fittest'.[50]

In Volume I of his book *Principles of Biology* published in 1864, philosopher and civil engineer Herbert Spencer had been the first to introduce the expression 'survival of the fittest'.[51] Now here was Darwin, humorously applying Spencer's expression to the matter in hand!

Chapter 18

Alfred Russel Wallace

As has already been pointed out, there were many similarities in make up between Wallace and Darwin.

A modest man with an enquiring mind

Referring to his life as a young man in his early twenties, Wallace declared:

> I do not think that at this time I could be said to have shown special superiority in any of the higher mental faculties, but I possessed a strong desire to know the causes of things … . If I had one distinct faculty more prominent than another, it was the power of correct reasoning from a review of the known facts in any case to the causes or laws which produced them, and also in detecting fallacies in the reasoning of other persons.[1]

A disciplined scientist: his powers of observation: a great collector

Having returned to London from the Far East in the spring of 1862, Wallace, employing a methodology similar to that of Darwin in the Galapagos Islands, described how he

> had determined to keep a complete set of certain groups [of 'birds, butterflies, beetles, and land-shells'] from every island or district locality which I visited for my own study on my return home, as I felt sure they would afford me very valuable materials for working out the geographical distribution of animals in the [Malay] archipelago …[2]
>
> During the succeeding five years I continued the study of my collections, writing many papers … five or six on the special applications of the theory of natural selection.[3]

Just as, at the Galapagos, Darwin had described how particular species of birds varied from island to island, so Wallace said of the butterflies of Malaya,

> the family presents us with examples of differences of size, form, and colour, characteristic of certain localities, which are among the most singular and mysterious phenomena known to naturalists.[4]

Authorship

Wallace's literary output was nothing like as prodigious as that of Darwin but, nevertheless, he wrote many books, including *The Geographical Distribution of Animals* (1876), *Travels on the Amazon and Rio Negro* (1889) and *Man's Place in the Universe* (a volume about astronomy, 1903).

His generous attitude towards Darwin as a fellow scientist

On 6 October 1858 Wallace, demonstrating a degree of generosity and selflessness to rival that of Darwin himself, told Hooker,

> I … look upon it as a most fortunate circumstance that I had a short time ago commenced a correspondence with Mr Darwin on the subject of 'Varieties', since it has led to the earlier publication of a portion of his researches & has secured to him a claim to priority which an independent publication either by myself or some other party might have injuriously affected … .
>
> It would have caused me much pain & regret had Mr Darwin's excess of generosity led him to make public my paper unaccompanied by his own …[5]

As already indicated Darwin was equally generous in respect of Wallace's achievements, saying,

> I hope it is a satisfaction to you to reflect – & very few things in my life have been more satisfactory to me – that we have never felt any jealousy towards each other, though [we are] in one sense rivals. I believe that I can say this of myself with truth, & I am absolutely sure that it is true of you.[6]

Did Darwin and Wallace ever meet? The answer is yes – at Down House on 12 September 1868, and in London on numerous occasions when Darwin visited his brother Erasmus.[7]

His colleagues

As with Darwin, Wallace included amongst the 'scientific friends … with whom I became most intimate', Lubbock (Darwin's neighbour at Downe, about whom more will be said shortly), Hooker, and Francis Galton,[8] but especially Huxley, who he said was 'as kind and genial a friend and companion as Darwin himself'. [9]

Compassion towards his fellow human beings

Of the eight-year-long expedition which he undertook to the Malay archipelago (commencing in 1854), Wallace declared

> The more I see of uncivilized people, the better I think of human nature on the whole, and the essential differences between civilized and savage man seem to disappear. [And of the Chinese of Malaya] … the great majority of them are quiet, honest, decent sort of people.[10]

As for colonialism, Wallace evidently regarded it with a benign resignation, as is indicated in a letter which he wrote to George Silk, his friend from childhood, from East Sumatra (Sumatra being an Indonesian island then administered by the Dutch East India Company).

> Personally, I do not much like the Dutch out here, or the Dutch officials; but I cannot help bearing witness to the excellence of their government of native races, gentle yet firm, treating their manners, customs, and prejudices with respect, yet introducing everywhere European law, order and industry.[11]

Politics: socialism

It was not Darwin's habit to pontificate upon matters political, even though, as has been demonstrated, he abhorred social injustice and, in particular, the practice of slavery. Wallace, however, was prepared to 'nail his colours to the mast'.

In 1837 the thirteen-year-old Wallace had been sent to London to reside with a master builder to whom his brother John was apprenticed as land surveyor. Here, he was introduced to the 'Hall of Science' in Tottenham Court Road – 'a kind of club or mechanics' institute for advanced thinkers among workmen, and especially for the followers of Robert Owen, the founder of the socialist movement in England'. (Socialism is defined as a political and economic theory of social organization which advocates that the means of production, distribution, and exchange should be owned or regulated by the community as a whole.[12]) In later life Wallace was to assert that he was 'absolutely convinced' as to the merits of socialism, which was 'the only form of society worthy of civilized beings, and that it alone can secure for mankind continuous mental and moral advancement …'.[13]

God and religion

Whereas Darwin did not entirely dismiss the possibility of the existence of a god of sorts – or at any rate, a creator – Wallace was to strike out in an entirely different direction, writing,

> my early home training was in a thoroughly religious but by no means rigid family where, however, no religious doubts were ever expressed, and where the word 'atheist' was used with bated breath as pertaining to a being too debased almost for human society.[14]

And he described his father as 'very religious in the orthodox Church of England way, and with such a reliance on Providence as almost to amount to fatalism'.[15]

However, at the age of thirteen, he read a book or paper (unidentified) which posed the following questions:

> Is God able to prevent evil but not willing? Then he is not benevolent. Is he willing but not able? Then he is not omnipotent. Is he both able and willing? Whence then is evil?

This, said Wallace, 'struck me very much, and it seemed quite unanswerable ....'[16] As for 'the horrible doctrine of eternal punishment as then commonly taught from thousands of pulpits by both the Church of England and Dissenters ...' this, in Wallace's view, was both

> degrading and hideous, and ... the only true and wholly beneficial religion was that which inculcated the service of humanity, and whose only dogma was the brotherhood of man.[17]
>
> As might have been expected, therefore, what little religious belief I had very quickly vanished under the influence of philosophical or scientific scepticism. This came first upon me when I spent a month in London with my brother John ...; and during the seven years I lived with my brother William, though the subject of religion was not often mentioned, there was a pervading spirit of scepticism, or free-thought as it was then called, which strengthened and confirmed my doubts as to the truth or value of all ordinary religious teaching.[18]

Having heard a Unitarian minister lecture on the subject of a book by German theologian David F. Strauss, entitled *The Life of Jesus Critically Examined* (published in 1835), Wallace declared:

> The now well-known argument, that all the miracles related in the Gospels were mere myths, which in periods of ignorance and credulity always grow up around all great men, and especially around all great moral teachers when the actual witnesses of his career are gone and his disciples begin to write about him, was set forth with great skill.[19]

Wallace and spiritualism

Spiritualism is defined as a system of belief or religious practice based on supposed communication with the spirits of the dead, especially through mediums – a medium being defined as a person claiming to be in contact with the spirits of the dead, and to be able to communicate between the dead and the living.[20] It will be recalled that

Darwin was thoroughly unimpressed by spiritualism, but not so Wallace, who came to the subject late in life.

Speaking of the 'materializations' (whereby a ghost, spirit, or similar entity appears in bodily form[21]) which he had observed at various seances (meetings at which people attempt to make contact with the dead, especially through the agency of a medium)[22] which he had attended, Wallace declared,

> the phenomena and the effect they produced upon me are fully described in the 'Notes of Personal Evidence', in my book on *Miracles and Modern Spiritualism* [published in 1874], ...[23] I had numerous opportunities of seeing phenomena with other mediums in various private houses in London.

These mediums included 'Mrs Marshall and her daughter-in-law', whom Wallace described as 'two of the best public mediums for physical phenomena I have ever met with ...'.[24]

Such phenomena, defined as a fact or situation that is observed to exist or happen, especially one whose cause or explanation is in question,[25] which, according to Wallace, 'materialized' before his very eyes, included a 'stately East Indian figure in white robes'; 'three female figures [which] appeared together'; 'a female figure with a baby'; and 'an Indian chief in war-paint and feathers'.[26]

In 1866, said Wallace, 'I wrote a pamphlet entitled "The Scientific Aspects of the Supernatural", which I distributed amongst my friends'.[27] Furthermore, in May/June 1874, an article by Wallace, entitled 'A Defence of Modern Spiritualism' appeared in the *Fortnightly Review*.[28] Wallace believed that

> there was always an unknown intelligence behind the phenomena – an intelligence that showed a human character and individuality, and an individuality which almost invariably *claimed* to be that of some person who had lived on earth, and who, in many cases, was able to prove his or her identity.[29]
>
> I feel myself that my character has continuously improved, and that this is owing chiefly to the teaching of spiritualism, that we are in every act and thought of our lives here building up a character which will largely determine our happiness or misery [in the life] hereafter; and also, that we obtain the greatest happiness ourselves by doing all we can to make those around us happy.[30]

Wallace and mesmerism

Wallace described how, during his time in London, he made the acquaintance of a dentist Dr Theodosius Purland, who 'was a very powerful mesmerist'. (Mesmerism is a therapeutic system devised by Austrian physician Franz A. Mesmer (1734–1815), who believed that 'magnetism' could be used to cure diseases.) This so impressed Wallace that when, in 1849, physician Dr John Elliotson and others founded the Mesmeric Infirmary, Weymouth Street, London for the treatment of Epilepsy, Deafness, Rheumatism, and other diseases, Wallace gave them his support.[31] Wallace's book *Miracles and Modern Spiritualism*, was published in 1874.

Phrenology

Wallace also studied phrenology – the detailed study of the shape and size of the cranium as a supposed indicator of character and mental abilities.[32]

The origin of life

Wallace pondered over how it was that life on Earth began and wrote:

> Soon after my return home [from the East], in 1862, [Henry W.] Bates and I, having both read [*First Principles of a New System of Philosophy*, published in 1862] and been immensely impressed by it, went together to call on Herbert Spencer … [its author]. Our thoughts were full of the great unsolved problem of the origin of life – a problem which Darwin's *Origin of Species* left in as much obscurity as ever – and we looked to Spencer as the one man living who could give us some clue to it … . But … our hopes were dashed at once. That, he said, was too fundamental a problem to even think of solving at present.[33]

As for Darwin, he told Wallace that if it could be shown that life had generated itself spontaneously (i.e. rather than having been created), then this 'would be a discovery of transcendent importance'.[34] In other words, if it was proved that God was not necessary, even in the role of creator, then this would have troubled Darwin not one iota.

Was man continuing to evolve?

Wallace's *Contributions to the Theory of Natural Selection* (published in 1864), contained an article on 'The Development of Human Races under the Law of Natural Selection', of which 'the most original and important part' was

> that in which I showed that so soon as man's intellect and physical structure led him to use fire, to make tools, to grow food, to domesticate animals, to use clothing, to build houses, the action of natural selection was diverted from his body to his mind, and thenceforth his physical form remained stable while his mental faculties improved. My paper shows why … the form and structure of our body is permanent, and that it is really the highest type now possible on the earth. The fact that we have not improved physically over the ancient

> Greeks, and that most savage races – even some of the lowest in material civilization – possess the human form in its fullest symmetry and perfection, affords evidence that my theory is the true one.[35]

What Wallace did not here take into account was that two to three millennia is an insignificant time period, when compared with the vastness of the evolutionary timescale.

Where Wallace differed from Darwin in matters scientific

In his autobiography Wallace summarized 'the four chief points' on which he 'differed from Darwin'. Unlike Darwin, he

> 1. believed that some agency other than natural selection, and analogous to that which first produced organic life, had brought into being his [Man's] moral and intellectual qualities.
>
> 2. Darwin believed that in the case of certain animals the males had obtained their bright colours, or other ornaments, by selection through female choice.

In other words, the sexual selection of brightly-coloured males by the females ensured that such characteristics were preserved (birds being known to have colour vision).

> I, on the other hand, believe that natural selection had operated independently on the two sexes, and each had acquired colouration or form according to its need for protection. The females, being often more exposed to danger than the males (as in the case of sitting birds), had acquired [i.e. evolved] more subdued colouration whilst the males had remained bright and comparatively conspicuous.
>
> 3. Darwin thought that the arctic plants found on isolated mountain tops within the tropics could only be explained by the spreading of the arctic flora over the tropics during the glacial period. From a study of the flora of oceanic islands, I had come to the conclusion that the mountain flora had been derived by aerial transmission of seeds either by birds or by gales.
>
> 4. Darwin always believed in the inheritance of acquired characteristics, such as the results of use or disuse of organs, and the effects of climate, food, etc., on the individual. I also accepted this theory at first, but when I had studied Mr [Francis] Galton's experiments and [German biologist] Dr [F. L. August] Weismann's theory of the continuity of the germ-plasm I had to change my views.[36]

(According to Weismann's 'germ plasm theory', in multicellular organisms only the 'germ cells' – 'gametes' – i.e. the cells of the ova and sperm, contribute to inheritance).

Eugenics
In his later years Wallace contributed to the Eugenics debate, as will be seen.

Chapter 19

Variation: The Theory of Pangenesis

On 30 January 1868 Darwin's book *The Variation of Animals and Plants under Domestication* was published by John Murray. The 'object of this work', he said, was to show 'the amount and nature of the changes which animals and plants have undergone whilst under man's dominion, or which bear on the general principles of variation'.[1]

There was no doubt that variation (or variability) was a feature of the natural world. The question was, how do variations come about? But first, Darwin turned his attention to the origin of life itself. Reiterating what he had said on a previous occasion, he declared that not only was it possible to

> conclude that at least all the members of the same class have descended from a single ancestor … [but also] as the members of quite distinct classes have something in common in structure and much in common in constitution, analogy would lead us one step further, and to infer as probable that all living creatures are descended from a single prototype.[2]

However, 'the first origin of life on this earth, as well as the continued life of each individual, is at present quite beyond the scope of science … .'[3]

Darwin declared that 'variability … mainly depends on changed conditions of life', but he confessed that it was 'governed by infinitely complex and unknown laws'.[4] Elaborating on this further, he declared that:

> Changes of any kind in the conditions of life, even extremely slight changes, often suffice to cause variability. Excessive of nutriment is perhaps the most efficient single exciting [i.e. that which brings out or give rise to[5]] cause.[6] We have reason to suspect that an habitual excess of highly nutritious food, or an excess relatively to the wear and tear of the organisation from exercise [i.e. excessive exertion], is a powerful exciting cause of variability.[7]

(Today, in the twenty-first century and with the advancement of science, it is becoming increasingly apparent that Darwin's views about the role which factors external to the body – such as food, exercise, the environment, etc., play in producing variations are not as far-fetched as might once have been be supposed, as will shortly be seen.)

It was also Darwin's view that 'Variation often depends … on the reproductive organs being injuriously affected by changed conditions … .'[8] These notions were 'shots in the dark' on the part of Darwin, who admitted as much by observing that

> certain extraordinary peculiarities have … appeared in a single individual out of many millions, all exposed in the same country to the same general conditions of life … [This led him] to conclude that such peculiarities are not directly due to the action of the surrounding conditions, but to unknown laws acting on the organisation or constitution of the individual … .[9]

Finally, in a letter to Wallace dated 22 November 1870, Darwin appears resigned to the fact that, in his words, 'we know nothing about [the] precise cause of each variation'.[10] (In this particular case he was referring to changes in colour of butterflies with succeeding generations.)

The truth was that, in his efforts to explain 'variation', Darwin was attempting the impossible, for the science of this subject was as yet in its infancy.

The inheritance of variations

Wrote Darwin

> When a new peculiarity ['variation'] first appears, we can never predict whether it will be inherited. If both parents from their birth present the same peculiarity, the probability is strong that it will be transmitted to at least some of their offspring.[11]

Here, Darwin was reflecting the fact that some genes are 'dominant' in nature, whilst others are 'recessive'. (Gene is a term coined in 1909 by Danish botanist Wilhelm Johannsen and now defined as a unit of heredity that is transferred from a parent to offspring and is held to determine some characteristic of the offspring.[12]) In the meantime, to him, the fact that not every feature of the parents was inherited by the offspring (what Darwin referred to as 'non-inheritance') was 'intelligible [only] on the principle, that a strong tendency to inheritance does exist, but that it is overborne by hostile or unfavourable conditions of life'.[13]

Darwin also declared, without supporting evidence and in a manner reminiscent of Lamarck, that 'the increased use of a muscle with its various detached parts, and the increased activity of a gland or other organ, lead to their increased development'. On the other hand 'disuse has a contrary effect' and 'a part becomes diminished by disuse for [i.e. over] many generations … '.[14]

> I do not believe any other person has taken such pain to show that the effects of use & disuse are inherited, as I have done.[15]

Other questions which occupied Darwin's mind were:

a. What was it that determined which characteristics were inherited when two different breeds were crossed (e.g. grey mice and white mice)?

b. Why it was that 'long-continued close interbreeding between the nearest relations diminishes the constitutional vigour, size, and fertility of the offspring; and occasionally leads to malformations …?'[16]

c. How may a characteristic 'of so grave a nature as to deserve to be called a monstrosity [a grossly malformed animal, plant, or person]'[17] suddenly present itself?[18]

d. 'How it was possible for a character possessed by some remote ancestor suddenly to reappear in the offspring … '[19] – a phenomenon that he termed 'reversion',[20] meaning a return to a previous state.

Natural selection

According to Darwin 'species have generally originated by the natural selection of extremely slight differences',[21] and

> each slight modification of structure which was in any way beneficial under excessively complex conditions of life has been preserved, whilst each which was in any way injurious has been rigorously destroyed. And the long-continued accumulation of beneficial variations will infallibly have led to structures as diversified, as beautifully adapted for various purposes and as excellently co-ordinated, as we see in the animals and plants around us.[22]

This, in a nutshell, summarized what was at the heart of Darwin's great theory, and it was this that marked him out (along with Wallace) as one of the greatest original thinkers of all time.

However, said Darwin, some naturalists 'will never admit that one natural species has given birth to another until they behold all the transitional steps'.[23]

The selective breeding of animals under domestication

Animals that he included in his study were dogs, cats, horses, asses, pigs, cattle, sheep, goats, rabbits, pigeons, fowl, ducks, geese, peacocks, turkeys, guinea fowl, canaries, goldfish, bees, and moths.

> Although man does not cause variability and cannot even prevent it, he can select, preserve, and accumulate the variations given to him by the hand of nature almost in any way which he chooses; and thus he can certainly produce a great result.[24] It can … be clearly shown that man …, by preserving in each successive generation the individual [animal or plant] which he prizes most, and by destroying the worthless individuals, slowly, though surely, induces great changes.[25]

On the practical and proactive side, wrote Darwin,

> When a man attends rather more closely than is usual to the breeding of his animals, he is almost sure to improve them to a slight extent.[26] [However] as a consequence of continued variability, and more especially of reversion, all highly improved races, if neglected or not subjected to incessant selection, soon degenerate.[27]

Darwin gave as examples of species that could be 'improved': the racehorse, in terms of its fleetness; and livestock, in order to produce prize cattle and sheep. He concluded, 'There can be no doubt that methodical selection has effected and will effect wonderful results.'[28] But, he asked, was there any limit as to the 'amount of variation in any part or quality' which could be achieved? For instance, was it possible to produce gooseberries of ever-increasing weight; beetroot which yielded 'a greater percentage of sugar'; or wheat and other types of grain, which 'produce heavier crops than our present varieties'?[29] With an ever-increasing world population these branches of science are, of course, of even greater relevance today.

Man

Finally, in the chapter entitled 'Inheritance', Darwin turned his attention to mankind, but not in the great detail which the world had hoped for and anticipated of him. Instead, he was content to observe that in human beings, disorders such as insanity, epilepsy, myopia, squint, hypermetropia, and polydactylism (the presence of supernumerary fingers and/or toes) were sometimes inherited. However, not all inherited characteristics were 'evil', and it was 'fortunate that good health, vigour, and longevity are equally inherited'.[30]

Pangenesis

In an attempt to explain the phenomenon of variation/variability, Darwin put forward the theory of *pangenesis*. This, however, was not a new theory for, as he himself admitted, 'views in many respects similar' to his had previously 'been propounded by various authors'.[31]

The theory of pangenesis stipulates that 'every separate unit or cell of an organism reproduces itself by contributing its share to the germ or bud of the future off-spring'.[32] According to Darwin, pangenesis worked in the following way:

i) Reproduction

> ovules, spermatozoa, and pollen grains, the fertilized egg or seed, as well as buds, include and consist of a multitude of germs thrown off from each separate part or unit [of the body].[33]

These germs he called 'gemmules … the number and minuteness of which must be something inconceivable'.[34] Gemmules

> are dispersed throughout the whole system [of the living organism, and] when supplied with proper nutriment, [they] multiply by self-division, and are ultimately developed into units like those from which they were originally derived. They are collected from all parts of the system to constitute the sexual elements, and their development in the next generation forms a new being … . Hence, it is not the reproductive organs or buds which generate new organisms, but the units of which each [parent] individual is composed.[35]

ii) Variation

> When two forms are crossed, one is not rarely found to be prepotent in the transmission of its characters over the other; and this we can explain by assuming … that the one form has some advantage over the other in the number, vigour, or affinity of its gemmules.[36]

iii) Divergence

> The crossing of distinct forms, which have already become variable, increase in the offspring the tendency to further variability by the unequal commingling of the characters of the two parents, by the reappearance of long-lost characters, and by the appearance of absolutely new characters.[37]

iv) Healing

> Referring to the healing process, Darwin declared that it was impossible to decide 'whether the ordinary wear and tear of the tissues is made good by means of gemmules, or merely by the proliferation of pre-existing cells'.[38]

v) Rudimentary or 'vestigial' organs (i.e. those which may have once had a purpose but now no longer do so).

Darwin declared that 'gemmules derived from reduced and useless parts would be more likely to perish than those freshly derived from other parts which are still in full functional activity'.[39]

vi) Reversion

> Darwin postulated that 'all organic beings … include [i.e. contain] many dormant gemmules derived from their grandparents and more remote progenitors, but not from all their progenitors'.[40] 'Reversion depends on the transmission from the forefather to his descendants of dormant gemmules, which occasionally become developed under certain known or unknown conditions.'[41]

Here Darwin is attempting to explain the commonly observed phenomenon that certain characteristics may skip one or more generations, only to reappear at a later stage.

vii) Inherited disease

> Each animal or plant may be compared with a bed of soil full of seeds, some of which soon germinate, some lie dormant for a period, whilst others perish. When we hear it said that a man carries in his constitution the seeds of an inherited disease, there is much truth in the expression.[42]

Here, Darwin appears to be making an analogy between seeds and gemmules.

In contrast to the theory of evolution, as propounded jointly by Darwin and Wallace, the theory of pangenesis lapsed into obscurity and was eventually forgotten. The science of the time was not sufficiently advanced for it to be proved or disproved, and when scientific knowledge *did* progress, the 'gemmule' was found to be nothing more than an imaginary concept, with no basis in reality. However, Darwin must be applauded for highlighting those aspects of reproduction, inheritance, and the capacity of the body to heal, for which there was, as yet, no explanation.

* * *

To physician and naturalist William Ogle, Darwin wrote on 6 March 1868 to say:

> I thank you most sincerely for your letter which is very interesting to me. [This letter has not been traced] I wish I had known of these views of Hippocrates before I had published [*Variation*], for they seem almost identical with mine … .

> The whole case is a good illustration of how rarely anything is new. The notion of pangenesis has been a wonderful relief to my mind, for during long years I could not conceive any possible explanation of inheritance, of development &c &c, or understand in the least in what reproduction by seeds & buds consisted.[43]

This was a reference to Greek physician Hippocrates (c.460–c.377BC), who

> propounded a theory according to which minute particles from every part of the body entered the seminal substance [or, in modern parlance, the spermatozoa and ovules] of the parents, and by their fusion gave rise to a new individual exhibiting the traits of both of them.[44]

To Julius V. Carus, German comparative anatomist, Darwin wrote on 21 March 1868 to say, in respect of the theory of pangenesis:

> All cases of inheritance & reversion & development now appear to me under a new light; whether this is false or true.[45]

In other words, Darwin was aware that his pangenesis theory was just a theory and not a scientific fact.

* * *

In that year of 1868 the indefatigable Darwin was to be found researching for another book which would be entitled *The Descent of Man, and Selection in Relation to Sex* and to this end, on 18 March, he wrote to entomologist Henry W. Bates, as follows:

> It has occurred to me that you must occasionally come across Missionaries or dealers [traders] who have long lived intimately with Savages; in this case, if you can, oblige me by leading conversation [i.e. let me know your views as] to the notion of savages about the beauty of women, & secondly & *more especially* how far the women have any indirect influence in getting men, whom they prefer or admire, to court them or purchase them from their parents.[46]

In other words Darwin wished to know what factors in male and female 'savages' influence their choice of mate.

On 18 April Darwin wrote to John J. Weir to say, in respect of his friend and colleague Alfred Russel Wallace, 'I always distrust myself when I differ from him ….'[47]

Hooker informed Darwin on 16 June that he had attended the fifth triennial festival in honour of George Frideric Handel at London's Crystal Palace, where he had heard a performance of the composer's oratorio 'Messiah'.[48] To this, Darwin replied on 17 June:

> I am glad you were at the Messiah: it is the one thing that I sh^{d} like to hear again, but I daresay I sh^{d} find my soul too dried up to appreciate it, as in old days; & then I sh^{d} feel very flat, for it is a horrid bore to feel, as I constantly do, that I am a withered leaf for every subject except science. It sometimes makes me hate science, though God knows I ought to be thankful for such a perennial interest which makes me forget for some hours every day my accursed stomach.[49]

Here was an indication that Darwin was suffering from a chronic medical condition, the nature of which will be discussed shortly.

Wallace wrote to Darwin on 30 August to say of his [Wallace's] attendance at a recent meeting of the British Association for the Advancement of Science, held in Norwich, 'Darwinianism was in the ascendant at Norwich; (I hope you do not dislike the word, for we really *must* use it,) … .'[50]

In early 1869 Darwin was busy preparing a new edition of *Origin*. That October he wrote to the US natural historian James Orton to say, 'Although I have never had any quarrel with Prof. [Richard] Owen, he has used such language about me that I can hold no communication with him.'[51]

Chapter 20

Sir Francis Galton

Francis Galton, anthropologist, explorer, and geographer, born in 1822, was the son of Samuel T. Galton (a banker), and Frances A. V. Galton (née Darwin). He and Darwin, both being grandsons of Erasmus Darwin, Charles as a result of Erasmus's first marriage to Mary, and Francis as a result of Erasmus's second marriage to Elizabeth, were, therefore, half-cousins.

Galton's *Hereditary Genius: an Enquiry into its Laws and Consequences* was published in 1869, a decade after Darwin's *The Origin of Species*. By conducting genealogical research into 'no less than 300 families containing between them nearly 1,000 eminent men …',[1] Galton, in his words, had endeavoured to find evidence of 'hereditary genius'. These families he divided into groups, which included judges, statesmen, peers of the realm, military commanders, literary men, men of science, poets, musicians, 'divines' (senior clerics), senior classicists from Cambridge University, oarsmen, and wrestlers. Then, using a sliding scale, he classified the individuals in various groups 'according to their natural gifts'. His conclusions were as follows:

> i. The nearer kinsmen of the eminent Statesmen were far more rich in ability than the more remote.[2] [This also applied to judges.]
>
> ii. More than one half of the great literary men … had kinsmen of high ability.[3]
>
> iii. At least 40 per cent of the [fifty-six] Poets [in the survey] … had eminently gifted relations.[4]
>
> iv. Of the twenty-six musicians whom he studied about 1 in 5 … had eminent kinsmen…[5]
>
> v. Of the forty-two illustrious ancient painters [whom he studied, about half of them he described as] possessing eminent relations.[6]

However, instead of using objective criteria for assessing the abilities of members of the various groups, Galton assumed that their rank alone was sufficient proof of their prowess. For example, of judges he asserted that 'the office of a judge is really a sufficient guarantee that its possessor is exceptionally gifted'.[7] As for statesmen, he declared that 'as is the case in every other profession, none, except those who are extraordinarily and peculiarly gifted, are likely to succeed in parliamentary life ...'.[8]

When he came to compare 'the worth of different races', Galton concluded that 'the number among the negroes of those whom we should call half-witted men, is very large.' 'The Australian type [i.e. the Aboriginal] is at least one grade below the African negro. The average standard of the Lowland [Scottish] and the English North-country men is decidedly a fraction of a grade superior to that of the ordinary English ... .' However, 'The ablest race of whom history bears record is unquestionably the ancient Greek ... .'[9] 'If we could raise the average standard of our race [presumably the English or the British] by only one grade, what vast changes would be produced!'[10]

However, the following passage by Galton appears to indicate that he laments the destruction of the weaker races.

> The number of the races of mankind that have been entirely destroyed under the pressure of the requirements of an incoming civilization, reads us a terrible lesson. Probably in no former period of the world has the destruction of the races of any animal whatever been effected over such wide areas and with such startling rapidity as in the case of savage man. In the North American Continent, in the West Indian Islands, in the Cape of Good Hope, in Australia, New Zealand, and Van Diemen's Land, the human denizens of vast regions have been entirely swept away in the short space of three centuries, less by the pressure of a stronger race than through the influence of a civilization they were incapable of supporting.[11]

But he goes on to argue that, for a variety of reasons, the 'savage' is incapable of being civilized.

> There is a most unusual unanimity in respect to the causes of incapacity of savages for civilization, among writers on those hunting and migratory nations who are brought into contact with advancing colonization, and perish, as they invariably do, by the contact. They tell us that the labour of such men is neither constant nor steady; that the love of a wandering, independent life prevents their settling anywhere to work, except for a short time, when urged by want and encouraged by kind treatment.[12]
>
> Much more alien to the genius of an enlightened civilization than the nomadic habit is the impulsive and uncontrolled nature of the savage. A civilized man must bear and forbear, he must keep before his mind the claims of the

> morrow as clearly as those of the passing minute; of the absent, as well as of the present. This is the most trying of the new conditions imposed on man by civilization, and the one that makes it hopeless for any but exceptional natures among savages to live under them. The instinct of a savage is admirably consonant with the needs of savage life; everyday he is in danger through transient causes; he lives from hand to mouth, in the hour and for the hour, without care for the past or forethought for the future: but such an instinct is utterly at fault in civilized life.[13]

Finally, Galton concludes that members of

> the human race were utter savages in the beginning; and that, after myriads of years of barbarism man has very recently found his way into the path of morality and civilization.[14]

There were several flaws in Dalton's arguments, one of which he himself admitted: that his sources of information were limited. Not only that, he made the mistake of judging so-called 'savages' by the so-called 'civilized' standards of Western societies. He discounted those particular skills which 'savages' themselves had evolved and, in so doing, failed entirely to recognize or value *their* cultures and religions.

Looking ahead, said Galton,

> The time may hereafter arrive, in far distant years, when the population of the earth shall be kept as strictly within the bounds of number and suitability of race, as the sheep on a well-ordered moor or the plants in an orchard-house; in the meantime, let us do what we can to encourage the multiplication of the races best fitted to invent and conform to a high and generous civilization, and not, out of a mistaken instinct of giving support to the weak, prevent the incoming of strong and hearty individuals.[15]

It was Galton who, in 1883, first coined the word *eugenic* – *eugenics* being defined as the science of improving a population by controlled breeding to increase the occurrence of desirable heritable characteristics.[16] He would develop his theme further in the years to come.

On 23 December 1869 Darwin wrote to Galton to say:

> I have only read about 50 pages of your Book … . I do not think I ever in all my life read anything more interesting & original. And how very well & clearly you put every point![17]

However, he would not commit himself as to whether or not he endorsed its sentiments.

In two years' time, Darwin would publish another book, this time related to the origin of man himself. Perhaps then he [Darwin] would give an opinion on this contentious subject.

To this, Galton, on the following day, declared that there was no one but Darwin

> whose approbation I prize more highly, on purely personal grounds, because I always think of you in the same way as converts from barbarism think of the teacher who first relieved them from the [intolerable] burden of their superstition. I used to be wretched under the weight of the old fashion 'arguments from design', of which I felt, though I was unable to prove to myself, the worthlessness.

('Argument from design' – that the apparent 'design' of the natural world is proof of the existence of God.)

> Consequently the appearance of your *Origin of Species* formed a real crisis in my life; your book drove away the constraint of my old superstition as if it had been a nightmare and was the first to give me freedom of thought.
>
> Believe me very sincerely yours Francis Galton[18]

Galton could not have written more succinctly and, in doing so, he echoed the sentiments of thinking people everywhere. It was as if, with the advent of Darwinism, a veil had been lifted, chains thrown off, bringing to mind, paradoxically, the Biblical quotation from the New Testament's First Book of Corinthians, 'For now we see through a glass, darkly; but then face to face ... .'[19]

* * *

In 1870 Oxford University awarded Darwin the honorary degree of Doctor of Civil Law (DCL). This was much to his surprise, for Oxford was the very place where the 'great evolution debate' had taken place. Darwin, however, was unable to accept the degree – which was awarded only in person – on the grounds that his ill health precluded him from attending the ceremony.

Darwin was awarded the Diploma of the Royal Academy of Science, Literature and Art of Belgium on 16 December of that year.

Chapter 21

The Descent of Man

When, in 1857, Alfred Russel Wallace asked Darwin whether he intended to 'discuss "man"' in his forthcoming book *The Origin of Species*, Darwin replied, 'I think I shall avoid [the] whole subject', on the grounds that it was too 'surrounded with prejudices'.[1] It would be another fourteen years before *The Descent of Man and Selection in Relation to Sex* was published on 24 February 1871. (It should be stressed that by 'Descent' Darwin meant 'Origin'.) What did it contain? How would it be received? What new controversies would it foment? Whatever the outcome, no one could deny that, like *Origin*, this was a serious work, based on painstaking research performed over many years.

In his Introduction, Darwin wrote:

> The sole object of this work is to consider, firstly, whether man, like any other species, is descended from some pre-existing form; secondly, the manner of his development; and thirdly, the value of the differences between the so-called races of man.[2]

Darwin points out 'the marvellous fact that the embryos of a man, dog, seal, bat, reptile &c., can, at first, hardly be distinguished from each other' and, he concludes,

> Thus we can understand how it has come to pass that man and all other vertebrate animals have been constructed on the same general model, why they pass through the same early stages of development, and why they retain certain rudiments in common. Consequently we ought frankly to admit their community [here, perhaps 'commonality' – meaning shared features or attributes [3] – might be a better word] of descent … .[4]

And declares,

> It is manifest that man is now subject to much variability. No two individuals [even] of the same race are quite alike.[5] [But] With respect to the causes of variability, we are in all cases very ignorant … .[6]

In respect of potential checks to population,

> With savages the difficulty of obtaining subsistence occasionally limits their number in a much more direct manner than with civilized people, for all tribes periodically suffer from severe famines. At such times savages are forced to devour much bad food, and their health can hardly fail to be injured.[7]

But 'what is probably the most important of all' of these checks is

> infanticide, especially of female infants, and the habit of procuring abortion. These practices now prevail in many quarters of the world … [and] appear to have originated in savages recognizing the difficulty, or rather the impossibility of supporting all the infants that are born.[8]

He compares the mental powers of man with that of the lower animals.

> Of all the faculties of the human mind, it will, I presume, be admitted that *Reason* stands at the summit. Only a few persons now dispute that animals possess some power of reasoning. Animals may constantly be seen to pause [in their actions], deliberate, and resolve.[9]

Under the heading, 'The moral sense …', Darwin reveals himself to be a champion of social justice, by his demand that slavery must be abolished. 'Although somewhat beneficial during ancient times', it 'is a great crime, yet it was not so regarded until quite recently, even by the most civilized nations'.[10]

> As man advances in civilization, and small tribes are united into larger communities, the simplest reason would tell each individual that he ought to extend his social instincts and sympathies to all the members of the same nation, though personally unknown to him. This point being once reached, there is only an artificial barrier to prevent his sympathies extending to the men of all nations and races.[11]

In other words Darwin, an ardent 'abolitionist', took the view that instead of enslaving people, the hand of friendship should be extended to them. However,

> The very idea of humanity, as far as I could observe [i.e. during the voyage of HMS *Beagle*] was new to most of the Gauchos of the Pampas. This virtue, one of the noblest with which man is endowed, seems to arise incidentally from our sympathies becoming more tender and more widely diffused, until they are

> extended to all sentient beings. As soon as this virtue is honoured and practised by some few men, it spreads through instruction and example to the young, and eventually becomes incorporated in public opinion.[12]

Darwin then turns his attention to the mind, in respect of man vis-à-vis the (other) animals.

> There can be no doubt that the difference between the mind [i.e. intellectual capacity] of the lowest man and that of the highest animal is immense. Nevertheless the difference in mind between man and the higher animals, great as it is, certainly is one of degree and not of kind. We have seen that the senses and intuitions, the various emotions and faculties, such as love, memory, attention, curiosity, imitation, reason, &c, of which man boasts, may be found in an incipient, or even sometimes in a well-developed condition, in the lower animals.[13]

These observations chime perfectly with Darwin's conviction that man is part of the animal kingdom. However,

> The moral sense perhaps affords the best and highest distinction between man and the lower animals … . I have so lately endeavoured to shew that the social instincts – the prime principle of man's moral constitution – with the aid of active intellectual powers and the effects of habit, naturally lead to the golden rule, 'As ye would that men should do to you, do ye [also] to them likewise' [a quotation from the Gospel of St Luke],[14] and this lies at the foundation of morality.[15]

Many years later Emma, as a devout Christian, was to take exception to Darwin's opinion 'that *all* morality has grown up by evolution'. The fact that her husband regarded morality as something which a), was based on instinct rather than on religion and b), had predated religion[16] was, she said, 'painful to me.'[17]

Under the heading 'On the Development of the Intellectual and Moral Faculties during Primeval and Civilized Times', Darwin declares:

> All that we know about savages, or may infer from their traditions and from old monuments, the history of which is quite forgotten by the present inhabitants, shew that from the remotest times successful tribes have supplanted other tribes. Relics of extinct or forgotten tribes have been discovered throughout the civilized regions of the earth, on the wild plains of America, and on the isolated islands in the Pacific Ocean. At the present day

> civilized nations are everywhere supplanting barbarous nations, excepting where the climate opposes [i.e presents] a deadly barrier; and they succeed mainly, though not exclusively, through their arts [i.e. skills], which are the products of the intellect. It is, therefore, highly probable that with mankind the intellectual faculties have been mainly and gradually perfected through natural selection … .[18]

This matter is a controversial one, for Darwin implies that the skills of so-called 'savage tribes' are inferior, whereas, in many cases these skills have enabled them to survive in some of the world's most hostile environments for centuries – if not millennia. It is also arguable, to say the least, as to whether or not, for example, the conquest of North America and the decimation of its indigenous peoples during the eighteenth and nineteenth centuries can be described as 'natural selection'.

Under the heading 'Natural Selection as affecting Civilized Nations', Darwin states that the following remarks of his are primarily based on the works of William Rathbone Greg (Unitarian, mill-owner and social commentator), A. R. Wallace and Francis Galton.

> With savages, the weak in body or mind are soon eliminated; and those that survive commonly exhibit a vigorous state of health. We civilized men, on the other hand, do our utmost to check the process of elimination; we build asylums for the imbecile, the maimed, and the sick; we institute poor-laws, and our medical men exert their utmost skill to save the life of every one to the last moment. There is reason to believe that vaccination has preserved thousands, who from a weak constitution would formerly have succumbed to small-pox. Thus the weak members of civilized societies propagate their kind. No one who has attended to the breeding of domestic animals will doubt that this must be highly injurious to the race of man. It is surprising how soon a want of care, or care wrongly directed, leads to the degeneration of a domestic race; but excepting in the case of man himself, hardly any one is so ignorant as to allow his worst animals to breed.

Now, at last, Darwin has come off the proverbial fence by implying that an over-zealous 'welfare state' may actually serve to set the evolutionary process into reverse. But then he presents an alternative point of view.

> The aid which we feel impelled to give to the helpless is mainly an incidental result of the instinct of sympathy…. Nor could we check our sympathy, even at the urging of hard reason, without deterioration in the noblest part of our nature.

In other words, to ignore the plight of the 'helpless' would be nothing short of ignoble.

> We must therefore bear the undoubtedly bad effects of the weak surviving and propagating their kind; but there appears to be at least one check in steady action, namely that the weaker and inferior members of society do not marry so freely as the sound; and this check might be indefinitely increased by the weak in body or mind refraining from marriage, though this is more to be hoped for than expected.[19]

Therefore, the 'weak' and the 'inferior' were to be encouraged, but not forcibly coerced to refrain from procreation.

But what of those who would today be called 'anti-social', and the danger that those of a 'superior class' might be overwhelmed by their numbers?

> A most important obstacle in civilized countries to an increase in the number of men of a superior class has been strongly insisted on by [essayist] Mr [William Rathbone] Greg and Mr Galton, namely, the fact that the very poor and reckless, who are often degraded by vice, almost invariably marry early, whilst the careful and frugal, who are generally otherwise virtuous, marry late in life, so that they may be able to support themselves and their children in comfort. Those who marry early produce within a given period not only a greater number of generations, but, as shewn by [Scottish physician] Dr [James Matthews] Duncan, they produce many more children. The children, moreover, that are born by mothers during the prime of life are heavier and larger, and therefore probably more vigorous, than those born at other periods. Thus the reckless, degraded, and often vicious members of society tend to increase at a quicker rate than the provident and generally virtuous members. In the eternal 'struggle for existence', it would be the inferior and less favoured race that had prevailed [i.e. would prevail] – and prevailed by virtue not of its good qualities but of its faults.[20]

Finally, Darwin agonizes, yet again, about the danger of overpopulation.

> It is impossible not to regret bitterly, but whether wisely is another question, the rate at which man tends to increase; for this leads in barbarous tribes to infanticide and many other evils, and in civilized nations to abject poverty … .[21]

He indicates that Huxley agrees with him, in believing man to be a member of the animal kingdom.

> Our great anatomist and philosopher, Prof. Huxley ... concludes that man in all parts of his organization differs less from the higher apes, than these do from the lower members from the same group. Consequently there 'is no justification for placing man in a distinct order'.[22]

(The divergence of humans from the ape family - from their common ancestor - is believed to have occurred sometime between 5 million and 7 million years ago.)

Remarks such as these would have infuriated the Anglican hierarchy; an immediate difficulty being that in the Book of Genesis it states, 'And God said, Let us make man in our image, after our likeness ... So God created man in his *own* image, in the image of God created he him...'.[23]

Equally, Darwin affirms that each of the various so-called 'human races', belongs to the human race as a whole.

> Although the existing races of man differ in many respects, as in colour, hair, shape of skull, proportions of the body, &c., yet if their whole structure be taken into consideration they are found to resemble each other closely in a multitude of points.[24]

Under the heading, 'On the Extinction of the Races of Man', Darwin declares that

> Man can long resist conditions which appear extremely unfavourable for his existence. Extinction follows chiefly from the competition of tribe with tribe, and race with race.

He then goes on to describe in detail how, when 'civilized nations' conquer 'barbarians' – i.e., the native inhabitants of 'uncivilized' countries – then there are many factors which cause the gradual extinction of the latter, 'the most potent of all the causes of extinction' being, 'in many cases ... lessened fertility and ill-health, especially amongst the children, arising from changed conditions of life ...'.

In other words, the native inhabitants are simply unable to adapt to the new way of life imposed upon them by the invader/colonist.[25]

In a chapter entitled 'Principles of Sexual Selection [defined as natural selection arising through preference by one sex for certain characteristics in individuals of the other sex]',[26] Darwin declares:

> Just as man can improve the breed of his game-cocks by the selection of those birds which are victorious in the cockpit, so it appears that the strongest and most vigorous males, or those provided with the best weapons, have prevailed under nature, and have led to the improvement of the natural breed or species.[27]

'Prevailed under nature' – agreed, but the notion that this has 'led to the improvement of the natural breed or species' is debatable, to say the least. For example, how does it square with the overthrow of supposedly civilized ancient Rome by those who Darwin would define as 'barbarians'?

On a personal level, according to Darwin,

> Our difficulty in regard to sexual selection lies in understanding how it is that the males which conquer other males, or those which prove the most attractive to the females, leave a greater number of offspring to inherit their superiority [i.e. their superior characteristics] than their beaten and less attractive rivals.[28]

The answer to this 'difficulty' is provided by Darwin in his 'Summary' to Chapter 18.

> The law of battle for the possession of the female appears to prevail throughout the whole great class of mammals. Most naturalists will admit that the greater size, strength, courage, and pugnacity of the male, his special weapons of offence, as well as his special means of defence, have been acquired or modified through that form of selection which I have called sexual. This does not depend on any superiority in the general struggle for life, but on certain individuals of one sex, generally the male, being successful in conquering other males, and leaving a larger number of offspring to inherit their superiority than do the less successful males.[29]

Finally, Darwin declares:

> In general we can only say that the cause of each slight variation and of each monstrosity lies much more in the constitution of the organism, than in the nature of the surrounding conditions; though new and changed conditions certainly play an important part in exciting organic changes of many kinds.[30]

Darwin therefore remained convinced that the environment did have a part to play in producing variations. He then proposed that, as with physical characteristics, 'It is not improbable that after long practice virtuous tendencies may [also] be inherited.'[31]

The way forward, according to Darwin

i. The importance of 'Sexual Selection'

In regard to 'sexual selection', Darwin observes, somewhat dryly, that

> Man scans with scrupulous care the character and pedigree of his horses, cattle, and dogs before he matches them; but when he comes to his own marriage he

> rarely, or never, takes any such care … . Yet he might by selection do something not only for the bodily constitution and frame of his offspring, but for their intellectual and moral qualities. Both sexes ought to refrain from marriage if they are in any marked degree inferior in body or mind … [However, he realized that] such hopes are Utopian and will never be even partially realized until the laws of inheritance are thoroughly known.[32]

ii. The necessity of making adequate provision for offspring

> The advancement of the welfare of mankind is a most intricate problem: all ought to refrain from marriage who cannot avoid abject poverty for their children …

iii. The 'most able' should have the most children

> The most able should not be prevented by laws or customs from succeeding best and rearing the largest number of offspring.[33]

Here Darwin appears to be implying that intellect and morality (as well as 'virtuous tendencies') are characteristics that are both desirable and heritable.

* * *

Darwin was, of course, aware that the publication of *The Descent of Man* would inevitably cause many 'hackles' to be raised, but nevertheless, he was not prepared to compromise. Said he, perhaps with a hint of mischief in his tone

> The main conclusion arrived at in this work, namely that man is descended from some lowly organized form, will, I regret to think, be highly distasteful to many. But there can hardly be a doubt that we are descended from barbarians.

He then proceeds to describe, in gory detail, the typical Fuegian whom he had encountered during the course of the *Beagle* voyage as

> a savage who delights to torture his enemies, offers up bloody sacrifices, practises infanticide without remorse, treats his wives like slaves, knows no decency, and is haunted by the grossest superstitions.[34]

Finally, he ends on a note of optimism.

> Man may be excused for feeling some pride at having risen, though not through his own exertions, to the very summit of the organic scale; and the fact of his having thus risen …, may give him hope for a still higher destiny in the distant future.

But then he brings the reader down to Earth with a bump.

> We must, however, acknowledge, as it seems to me, that man with all his noble qualities, with sympathy which [he] feels for the most debased, with benevolence which extends not only to other men but to the humblest living creature, with his god-like intellect which has penetrated into the movements and constitution of the solar system – with all these exalted powers – Man still bears in his bodily frame the indelible stamp of his lowly origin.[35]

Today, from the fossil record, it is believed that the last common ancestor from which Old World monkeys (*Cercopithecidae*) on the one hand, and lesser apes and great apes on the other, evolved, lived about 25 million years ago. From the great apes evolved orang-utans, gorillas, and the common ancestor of humans and chimpanzees. This common ancestor lived between 8 million and 6 million years ago, though its fossilized remains have yet to be discovered.[36]

Thus, spurred on by the support he had received from his followers throughout the world, Darwin had finally overcome his anxieties and reservations; gone public, and published *The Descent of Man*. Honestly and openly, he had voiced his hopes and fears for the human race, whilst at the same time insisting that, when it came to a head-to-head confrontation between scientifically based notions of how to improve the human race on the one hand, and humaneness on the other, then there was no contest – humaneness must always prevail.

* * *

Selfishness in nature: a view to which Darwin was adamantly opposed

Some time after April 1871, Darwin took issue with Frances J. Wedgwood, daughter of Emma's brother Hensleigh and his wife Frances, in respect of a communication which he had received from the former on the subject of evolution.

> I enter my protest against your making the struggle for existence (which is sufficiently melancholy fact) still more odious by calling it 'selfish competition'.

> A feline animal [which] is born rather bigger, fiercer or more cunning than others of the same or some other species & succeeds in life, & rears lots of savage little kittens, who get on very well in life, yet you cannot call this, even metaphorically, selfishness.[37]

An example would be a lion that wakes up one day feeling hungry. As a carnivore he is obliged to eat meat. For his survival it is essential for him to hunt down an antelope or a zebra. His behaviour is, therefore, instinctive; instinct being defined as an innate, typically fixed pattern of behaviour in animals in response to certain stimuli.[38] It might also be argued that other emotive words or phrases such as 'brutality', 'cruelty', and 'bitter struggle' are also inappropriate in respect of evolution in the natural world. Human beings are also animals, and even though they lead more complicated lives, when it comes to matters of survival, the same basic instincts exist.

Darwin was elected as 'Foreign Corresponding Member' of the Mathematics and Natural Sciences Section of the Imperial Academy of Science of Vienna on 1 August 1871. He described Hooker, on 20 October, as 'the best & oldest friend I have in the world'[39] and, on 7 December, was named 'Foreign Corresponding Member' of the Anthropological Society of Paris, and awarded the Society's Diploma. In the same month, when he wrote to his son Horace, he was pondering over why some men make discoveries whereas others do not.

> I have been speculating last night what makes a man a discoverer of undiscovered things; and a most perplexing problem it is. Many men who are very clever – much cleverer than the discoverers – never originate anything. As far as I can conjecture, the art consists in habitually searching for the causes or meaning of everything which occurs. This implies sharp observation & requires as much knowledge as possible of the subjects investigated.[40]

In November Scottish writer and dog breeder George Cupples, presented him with a deerhound puppy called 'Bran'.[41]

Chapter 22

Darwin and Freedom of Thought

To his lawyer Vernon Lushington, Darwin once wrote:

> On one point I have for many years vehemently [protested] in my own [mind], viz against the schemes of suggesting subject[s] for me to follow [i.e. study], I cannot conceive any scheme better adapted for stopping originality & great discoveries ....[1]

What constraints, if any, were imposed upon Darwin, in respect of his scientific researches and writings? First, it is necessary to consider what had gone before.

Galileo Galilei (1564–1642)

In 1632 the Italian physicist, mathematician, astronomer and philosopher Galileo, published his *Dialogue on the Two Principal Systems of the World*, in which he affirmed his support for Polish astronomer Nicolaus Copernicus (1473–1543) who proposed that the Earth and the planets revolve around the Sun and not vice-versa, as was formerly believed. For this, Galileo was tried as a so-called heretic (one who holds a belief or opinion contrary to orthodox [and] especially Christian doctrine[2]) by the Inquisition (*Inquisitio Haereticae Pravitatis* – Inquiry into Heretical Perversity, part of the Catholic Church's judicial apparatus). He was forced to recant and spent the remainder of his life under house arrest. Others suffered far worse fates than he.

Between the fifteenth and the nineteenth centuries, however, a series of events occurred which completely altered the environment in which scientists were obliged to work:

The Protestant Reformation

The Reformation – a sixteenth-century movement for the reform of abuses in the Roman Catholic Church ending in the establishment of the Reformed and Protestant Churches[3] – was set in train on 31 October 1517 by German religious reformer Martin Luther (1483–1546), Professor of Biblical Studies at Germany's University of

Wittenberg. In particular, Luther was disgusted by the venal greed of the clerics who grew wealthy by trading in 'indulgences' (an indulgence being a grant by the Pope of remission of the punishment in purgatory still due for sins after absolution, purgatory being, in Roman Catholic doctrine, a place or state of suffering inhabited by the souls of sinners who are expiating their sins before going to Heaven.)[4] Followers of Luther became known as Protestants.

Protestant Reformers took the view that neither the intercession of priests, departed saints, nor the Virgin Mary was necessary, as far as a believer's relationship with Christ was concerned. Luther was, therefore, anxious to circumvent the priesthood and bring the scriptures directly to the people. The only way in which this could be done was to translate the *Holy Bible* from the Greek into German, a feat which he himself achieved between the years 1522 and 1532. The outcome was that, in 1534, the so-called *Luther Bible* was printed in German at Wittenberg. (The first *Bible* in English was the one translated by English scholar William Tyndale; known as the '*Great Bible*', which was printed in 1539.) By his action, Luther had challenged the authority and jurisdiction of the Catholic Church.

King Henry VIII's break with Rome

Another challenge to Rome came in 1527 with the so-called English Reformation – the catalyst for which was the refusal of Pope Clement VII to annul King Henry VIII's marriage to Catherine of Aragon. The monarch reacted by rejecting the pontiff's authority and declaring himself (by the parliamentary Act of Supremacy of 1534) head of the Church of England.

The Civil War

When King Charles I of England (1600–49), who believed his Royal Prerogative – the 'rights and privileges exclusive to the sovereign'[5] – to be divinely ordained, came into conflict with Parliament, this led to the Civil Wars of 1642–49, at the conclusion of which the King was executed. According to historian Kenneth O. Morgan, this so-called English Revolution (which actually affected all three kingdoms) stood 'as a turning-point' and marked the beginning of a new 'age of pragmatism and individualism'.[6]

The Royal Society

The Royal Society of London for Improving Natural Knowledge was founded on 28 November 1660.

> The week after the formation of the new Society … word was brought that the King [Charles II] 'did well approve of it'. This support led to the Charter of Incorporation of July 1662 and to the Second Charter of April 1663, which is

> still the basis of the constitution of the Society. By these charters, Charles II and all later monarchs have been Patrons of the Society.
>
> The first fellows of the Royal Society were not scientists. They were courtiers, clergymen, doctors, lawyers, and merchants. They came together because they were fascinated by the possibility of a new way of thinking – a 'new philosophy' [which] relied on first-hand observation and experiment to understand the world. The new philosophy would be based on facts: facts that could be witnessed, discussed, verified and replicated. It was demanding and exciting, and very soon it would change the world.[7]

In his *History of the Royal Society* (published in 1667), English divine and man of letters Thomas Sprat, wrote of the society's early Fellows,

> one of the Principal Intentions they propos'd to accomplish, was a General Collection of all the Effects [belongings] of Arts, and the Common or Monstrous Works of Nature. This, they at first began by the casual Presents [bequests], which either Strangers, or any of their own Members bestowed upon them. And in short time it has increas'd so fast … that they have already drawn together in one Room, the greatest part of all the several kinds of things, that are scatter'd throughout the Universe.[8]
>
> The early Fellows combined their enthusiasm for collecting artefacts with a passion for record-keeping. They knew it would be vital to record the results of their experiments and observations so that later researchers could build on firm, factual evidence. At a time when little was known about the natural world, every scrap of new information was precious.
>
> From the outset, the Fellows realized the importance of publishing the results of their research. By publishing they could claim priority for new discoveries and inventions, and promote their work at home and abroad. In 1665 the Society's secretary, Henry Oldenburg, founded the *Philosophical Transactions* (the world's longest-running scientific journal).
>
> The Society also supported Fellows who wrote longer books, Isaac Newton's *Philosophiae Naturalis Principia Mathematica* [*The Mathematical Principles of Natural Philosophy*] being perhaps the most important scientific book ever published.[9]
>
> From the early decades of the eighteenth century, science began to be accepted by the British public as a useful and important activity. Science lectures were popular – some were delivered by Royal Society Fellows … .[10]
>
> In 1774 Josiah Wedgwood [I] [who was elected Fellow of the Royal Society in 1783] invented a close-grain ceramic material that he called 'jasper'. This enabled him to produce perfect portrait medallions quite cheaply. He produced

> portrait medallions of over seventy Fellows of the Royal Society, past and present, as well as many other British public figures.[11]

Darwin himself would one day receive the honour of being portrayed on such a Wedgwood plaque. (As already mentioned, Huxley, Lyell, Owen, FitzRoy, and Hooker were all Fellows of the Royal Society.)

> Until the early 19th century, Fellows of the Royal Society had been elected because they were interested in science, rather than necessarily being practitioners of science. This began to change as science became more established as a profession, and there were calls for reform. In 1847 the Society passed new statutes that restricted admission to men who had made distinguished contributions to science, or were active promoters of science.[12]

The reigning monarch in 1859, when Darwin's book *Origin* was published, was Queen Victoria. The monarch's subsidiary title (since 1521) was *Fidei Defensor* (Defender of the Faith) – which at that time meant the Catholic faith (which originally had been bestowed on Henry VIII by Pope Leo X in 1521 for his defence of the faith against the teachings of Luther). It is therefore paradoxical that under Her Majesty's patronage of the Royal Society, Darwin, a Fellow of that society since 1839, should be instrumental in disproving one of the key tenets of the Anglican faith, namely the Biblical account of Creation.

Amongst the possessions of the Royal Society are a bust of Darwin by the sculptor Horace Montford (who, incidentally, was born in Darwin's home town of Shrewsbury in Shropshire), and a portrait of Darwin by Mabel J. B. Messer (which, in turn, is a copy of a portrait of Darwin by John Collier, owned by the Linnaean Society).

The Enlightenment

The so-called Enlightenment, which heralded the 'Age of Reason', reached its zenith in the eighteenth century. It encouraged science and intellectual discourse, and opposed superstition, intolerance, and abuses by Church and state.

> Enlightenment thinkers were believers in social progress and in the liberating possibilities of rational and scientific knowledge. They were often critical of existing society and were hostile to religion, which they saw as keeping the human mind chained down by superstition.[13]

Leading Enlightenment figures were:

Frenchmen: Voltaire (writer and historian); Denis Diderot (writer); René Descartes (philosopher and mathematician); Jean Jacques Rousseau (political philosopher and educationist).

Dutchman: Baruch Spinoza (philosopher and theologian).
American: Benjamin Franklin (statesman, inventor and scientist).
Scotsmen: David Hume (philosopher and historian); Adam Smith (economist and philosopher); James Watt (engineer and inventor).
Irishman: Sir Richard Steele (essayist, dramatist and politician).

Meanwhile, England's principal Enlightenment figures were: Alexander Pope (poet); Mary Wollstonecraft (writer) and her husband William Godwin (political writer and novelist); John Locke (philosopher); Thomas Paine (radical political writer); Joseph Addison (essayist and politician); Daniel Defoe (writer and adventurer); Isaac Newton (scientist and mathematician); Matthew Boulton (engineer); Joseph Priestley (clergyman and chemist). However, what was most significant for Darwin was the fact that two members of his very own family were Enlightenment figures: namely, Josiah Wedgwood (I) and Erasmus Darwin (I).

This then, was the Darwinian milieu – the environment in which Darwin found himself – one in which, as a student of natural history, he was positively encouraged, rather than repressed, and one in which he knew that, even if his discoveries offended the Established Church, the worst that he could expect was its opprobrium. And if he offended the Establishment itself, then provided that he did not commit treason or foment insurrection – which was hardly likely – it was of no consequence. Meanwhile, he was not the first to have become fascinated by the question of the origin of species, mankind included; for many before him, including his grandfather Erasmus, had also wrestled with this great conundrum. Indeed, some have earned the right to be called 'proto-evolutionists'. Of these, Erasmus Darwin (I), Lamarck, Patrick Matthew, and William Charles Wells will now be discussed.

Chapter 23

Erasmus Darwin

Darwin's paternal grandfather, Erasmus, was born on 12 December 1731 at Elston Hall near Nottingham, the son of Robert Darwin, a lawyer of independent means, and his wife, Elizabeth (née Hill). Having studied classics and mathematics at St John's College, Cambridge, he went on to study medicine at Edinburgh, taking his MB degree (from Cambridge) in 1755. He then set up in medical practice, first in Nottingham and, subsequently in 1756, in Lichfield, Staffordshire.

In December 1757 Erasmus married Mary Howard who bore him five children, including Robert Waring Darwin, born in 1766 (the father of Charles Robert Darwin). In 1761 he was elected Fellow of the Royal Society. In 1770 Mary died after a long illness. Eleven years later, in 1781, Erasmus married Elizabeth Pole, a widow (who, allegedly, was the illegitimate daughter of Charles Colyer, 2nd Earl of Portmore), whereupon the couple relocated to Derbyshire.

Erasmus was a founder member of the Lunar Society – so called because its meetings were held on the afternoon of the Monday nearest to the time of the full moon. Its members included many of the great scientists and industrialists of the day, including clergyman and scientist Joseph Priestley, potter Josiah Wedgwood (I), and arms manufacturer Samuel Galton (grandfather of Sir Francis Galton). Many of these men were of the Unitarian faith which made them unpopular, and when they supported the French Revolution they became even more so.

In addition to pursuing his medical practice, the energetic Erasmus found time to experiment with gases and electricity; to study natural philosophy, chemistry, geology, and meteorology; and to design '[horse-drawn] carriages, a copying machine, and even a mechanical bird'![1] He created his own botanic garden and undertook the translation of the writings of Swedish naturalist Carl Linneaus (1707–78) and those of Linneaus's son Carl the younger (1741–83). Poems which Erasmus wrote included *The Loves of the Plants* and *The Economy of Vegetation*. Erasmus Darwin's *Zoonomia* was published in 1794 in two volumes, and it reveals that many of the questions which interested its author were the same ones as would one day exercise the mind of his grandson Charles.

***Zoonomia*, Volume I**

In a section entitled 'Generation' Erasmus reveals an interest in embryology and also in the mechanism of inheritance and heritable diseases.

The living filament

Erasmus postulated that:

> At the earliest period of its existence the [embryo], as secreted from the blood of the male, would seem to consist of a living filament with certain capabilities of irritation, sensation, volition, and association; and also with some acquired habits or propensities peculiar to the parent: the former of these are in common with other animals; the latter seem to distinguish or produce the kind of animal, whether man or quadruped, with the similarity of feature or form to the parent.[2]

As for this (supposed) living filament, its characteristics

> may have been gradually acquired during a million of generations, even from the infancy of the habitable earth … . [He also points out that this 'fact'] appears to have been shadowed or allegorized in the curious account in sacred writ [i.e. in Genesis, the first Book of the Old Testament] of the formation of Eve from a rib of Adam.
>
> From all these analogies I conclude, that the [embryo] is produced solely by the male, and that the female supplies it with a proper nidus [place of nurture], with sustenance, and with oxygenation ….

However, the constituency of the nidus 'may contribute to produce a difference in the form, solidity, and colour of the fetus …' and therefore, 'it follows that the fetus should so far [i.e. to some extent] resemble the mother'. Furthermore, 'This explains, why hereditary diseases may be derived either from the male or female parent … .

'From this account of reproduction', said Erasmus, 'it appears that all animals have a similar origin, viz. from a single living filament ….'

Selective breeding

Erasmus describes 'the great changes introduced into various animals by artificial [i.e. selective breeding by humans] or accidental cultivation', which had resulted in the appearance of stronger and faster horses; rabbits and pigeons of novel shape and character, etc. In fact, such animals 'have undergone so total a transformation, that we are now ignorant from what species of wild animals they had their origin.'

Charles Darwin.
(© English Heritage)

Charles Darwin. A chalk and watercolour drawing by George Richmond. (© English Heritage. By kind permission of Darwin Heirlooms Trust)

Jean Baptiste P. A. de Monet de Lamarck

Leonard Darwin

Robert FitzRoy, Captain of HMS *Beagle*

Alfred Russel Wallace

Francis Galton

Thomas Henry Huxley

Samuel Wilberforce, Bishop of Oxford

Emma Darwin (née Wedgwood). A chalk and watercolour drawing by George Richmond. (© English Heritage. By kind permission of Darwin Heirlooms Trust)

Richard Owen

Charles Lyell

Joseph Dalton Hooker

Thomas Robert Malthus

William Paley

John Locke

Josiah Wedgwood (II)

Samuel Butler

Adam Sedgewick

Robert Waring Darwin, from an engraving by Thomas Lupton after a painting by J. Pardon. (© English Heritage)

Susannah Darwin (née Wedgwood)

John Stevens Henslow

Down House in the 1870s

HMS *Beagle* in the Murray Narrows, Tierra del Fuego, by Conrad Martens. (© English Heritage. By kind permission of Darwin Heirlooms Trust)

Oxford University Museum. An engraving from the *Oxford Almanack* (1860). (Courtesy of the Oxford Museum of the History of Science)

Rhodnius prolixus of the family *Reduviidae*. (Courtesy of Toni Soriano Arandes)

An explanation for physical abnormalities, or even for the creation of a new species
'The great changes produced in the species of animals before their nativity,' said Erasmus, was due either to 'artificial cultivation' [i.e. selective breeding], to 'accident', or to 'the exuberance of nourishment supplied to the fetus' – which could result in the 'monstrous births' of creatures with additional limbs, claws, etc., or, conversely, to the absence of limbs or other physical features. It could even result in the appearance of 'a new species of animal'.

God as the creator of life
As to the origin of mammalian life, asked Erasmus,

> Would it be too bold to imagine, that in the great length of time, since the earth began to exist, perhaps millions of ages before the commencement of the history of mankind ..., that all warm blooded animals have arisen from one living filament, which THE GREAT FIRST CAUSE [i.e. Creator] endued with animality [an animal nature], with the power of acquiring new parts, attended with new propensities, directed by irritations, sensations, volitions, and associations; and thus possessing the faculty of continuing to improve by its own inherent activity, and of delivering down those improvements by generation to its posterity, world without end!

Divergence
The 'first link' of the 'perpetual chain of causes and effects [was riveted] to the throne of GOD ...', said Erasmus. Furthermore, this chain 'divides itself into innumerable diverging branches ...'. (This, of course, goes to the very heart of the principle of Divergence.)

Erasmus cites Linnaeus's early description of evolution – though this word was not in use at the time.

> Linnaeus supposes, in the Introduction to his Natural Orders, that very few [varieties of] vegetables were at first created, and that their numbers were increased by their intermarriages ... [and declares] This idea of the gradual formation and improvement [i.e. evolution] of the animal world seems not to have been unknown to the ancient philosophers.

This was a reference to Linnaeus's book *Systema naturæ per regna tria naturae, secundum classes, ordines, genera, species, cum characteribus, differentiis, synonymis, locis* or *System of nature through the three kingdoms of nature, according to classes, orders, genera and species, with characters, differences, synonyms, places*, which was first published in 1735.

God as the designer

Erasmus proceeds to criticize certain 'ancient philosophers, who contended that the world was formed from atoms …'. Had they

> ascribed their combinations to certain immutable properties received from the hand of the Creator … instead of ascribing them to a blind chance; the doctrine of atoms, as constituting or composing the material world by the variety of their combinations, so far from leading the mind to atheism, would strengthen the demonstration of the existence of a Deity, as the first cause of all things …

And he ends by quoting from the *Book of Common Prayer*, Psalm xix: *The heavens declare the glory of* GOD, *and the firmament sheweth his handywork. One day telleth another, and one night certifieth another; they have neither speech nor language, yet their voice is gone forth into all lands, and their words into the ends of the world.* He also quotes from Psalm civ: *Manifold are thy works,* O LORD! *in wisdom hast thou made them all.*[3]

***Zoonomia*, Volume II**

This is a comprehensive medical textbook, complete with suggested remedies for each disease mentioned. Said Darwin

> The book when published was extensively read by the medical men of the day, and the author was highly esteemed by them as a practitioner.[4]

However, as far as the opinions expressed by Erasmus in respect of the aetiology of the various diseases which he describes are concerned, these are largely uncorroborated and unsubstantiated.

In his biography of his grandfather (whom he never knew, and published in 1879), Darwin indicates just how enlightened and ahead of his time Erasmus was. For example, he was strongly in favour of prison reform; public health and hygiene; vaccination against smallpox; education; the abolition of slavery; and the humane treatment of the insane. Equally, he was strongly opposed to intemperance.[5] Erasmus was also a stickler for scientific rigour, as he indicates in 'The Botanic Garden' (a poem in two parts, published in 1791), where he laments the conjectural nature of the works of others.

> It may be proper here to apologize for many of the subsequent conjectures on some articles of natural philosophy, as not being supported by accurate investigation, or conclusive experiments.[6]

Said Darwin of his grandfather:

> He was what would now be called a liberal, or perhaps rather a radical. He seems to have wished for the success of the North American colonists in their war for independence … [and] Like so many other persons, he hailed the beginning of the French Revolution with joy and triumph.[7]
>
> The vividness of his imagination seems to have been one of his pre-eminent characteristics. This led to his great originality of thought.[8]

As for his religious beliefs, wrote Darwin, Erasmus

> has been frequently called an atheist, apparently as a convenient term of abuse; whereas in every one of his works distinct expressions may be found showing that he fully believed in God as the creator of the universe.[9]

Darwin does, however, admit that although Erasmus

> was certainly a theist in the ordinary acceptation of the term, he disbelieved in any revelation – the divine or supernatural disclosure to humans of something relating to human existence or the world[10]. Nor did he feel much respect for unitarianism, for he used to say that 'unitarianism was a feather-bed to catch a falling Christian'.[11]

What Erasmus meant by this was, presumably, that if a Christian, for one reason or another, lapses in his or her faith, then there was always Unitarianism to fall back on.

What were Darwin's views on *Zoonomia*?

In his autobiography Darwin describes how, during his time as a medical student at Edinburgh University, he took a walk with biologist Dr Robert Edmond Grant (from 1827, Professor of Comparative Anatomy at University College London), who

> burst forth in high admiration of Lamarck and his views on evolution. I listened in silent astonishment and as far as I can judge, and without any effect on my mind. I had previously read the *Zoonomia* of my [late] grandfather [Erasmus], in which similar views are maintained, but without producing any effect on me. At this time, I admired greatly the *Zoonomia*; but on reading it a second time, after an interval of ten or fifteen years, I was much disappointed; the proportion of speculation being so large to the facts given.[12]

In other words, Darwin was impressed neither by the arguments of Lamarck, nor by those of his own grandfather Erasmus, for as far as he was concerned, the scientific opinions of a family member were to be subjected to the same rigorous scrutiny as those put forward by others.

And yet Erasmus must be praised for his industriousness, and for his attempts to shed light on some of the unsolved conundrums relating to the natural world.

Erasmus died on 18 April 1802, seven years before his grandson Charles Darwin was born.

Chapter 24

Lamarck

Jean-Baptiste Pierre Antoine de Monet Chevalier de Lamarck was born in Bazentin, France in 1744. Like Darwin, he was to sample a number of different careers before finding his true vocation in life.

> He was destined by his father for an ecclesiastical career, and was entered as a student at the Jesuit College at Amiens. Yet he himself had no inclination to the calling desired by his father; and on the death of the latter in 1760, he made immediate use of his new liberty to leave the Jesuit College and join the French army … .[1]

However, having been forced to abandon the military, owing to ill health, he embarked on the study of medicine. But, once again, his life took a different turn when, through the influence of French political philosopher and educationist Jean-Jacques Rousseau, he became interested in botany and, in 1773, published his *Flore Française (French Flora)* in three volumes, which gained him entrance to the French Academy of Sciences.

In 1782 Lamarck became Keeper of the Herbarium at the Jardin du Roi in Paris. The herbarium was subsequently incorporated into the Muséum d'Histoire Naturelle when it was founded in 1793, and two chairs of botany were created, one of which was offered to Lamarck. He then became Professor of Invertebrate Zoology – even though he was a botanist of twenty-five years standing![2] In 1809, Lamarck's *Philosophie Zoologique* was published, of which he declared

> Experience in teaching has made me feel how useful a philosophical zoology would be at the present time. By this I mean a body of rules and principles, relative to the study of animals, and applicable to the other divisions of the natural sciences … . [To this end] I found myself compelled to consider the organization of the various known animals, to pay attention to the singular differences which it presents in those of each family, each order, and especially each class … .

Nonetheless, in his scientific researches, Lamarck makes it clear that he is determined to

> respect the decrees of that infinite wisdom [i.e. God] and confine myself to the sphere of a pure observer of nature. If I succeed in unravelling anything in her methods, I shall say without fear of error that it is has pleased the Author of nature to endow her with that faculty and power.[3]

In the event, however, rather than being a mere 'observer of nature', Lamarck finds that he cannot resist making deductions in respect of his scientific data, and when he does so, he ties himself in proverbial knots as he attempts to reconcile his observations with Biblical Creationist teachings – as will now be seen.

Evolution: how animals perfect themselves
Lamarck observed that

> in ascending the animals' scale, starting from the most imperfect animals, organization gradually increases in complexity in an extremely remarkable manner.
>
> I was greatly strengthened in this belief, moreover, when I recognized that in the simplest of all organisms there were no special organs whatever, and that the body had no special faculty but only those which are the property of all living things. As nature successively creates the different special organs, and thus builds up the animal organization, special functions arise to a corresponding degree, and in the most perfect animals these are numerous and highly developed.[4]

How was the increasing complexity of the animal kingdom achieved? Lamarck believed that he had the answer.

> Firstly, a number of known facts proves that the continued use of any organ leads to its development, strengthens it and even enlarges it, while permanent disuse of any organ is injurious to its development, causes it to deteriorate and ultimately disappear if the disuse continues for a long period through successive generations. Hence we may infer that when some change in the environment leads to a change of habit in some race of animals, the organs that are less used die away little by little, while those which are more used develop better, and acquire a vigour and size proportional to their use.[5]

In other words, environmental changes could produce variations, albeit indirectly, by first inducing a change in behaviour. Amongst the many examples quoted by Lamarck of how animals (allegedly) adapt their shapes and sizes to suit their environment, the example of the giraffe is one of the most graphic.

> This animal, the largest of the mammals, is known to live in the interior of Africa in places where the soil is nearly always arid and barren, so that it is obliged to browse on the leaves of trees and to make constant efforts to reach them. From this habit long maintained in all its race, it has resulted that the animal's fore-legs have become longer than its hind legs, and that its neck is lengthened to such a degree that the giraffe, without standing up on its hind legs, attains a height of six metres (nearly twenty feet).[6]

Attractive though Lamarck's theory may appear to be, he was unable to produce any scientific evidence to support it.

The spontaneous generation of life
Lamarck differentiates 'between the act of fertilization which prepares an embryo for the possession of life, and the act of nature which gives rise to direct generations …'.[7] And he believed that such 'spontaneous generations' occur in both the animal and plant kingdoms.[8]

Lamarck's views on variation
It was previously believed, said Lamarck, 'that every species is invariable and as old as nature, and that it was specially created by the Supreme Author of all existing things'. This was not the case, and he goes on to refer to 'the changes which nature is incessantly producing in every part without exception'.[9] However, echoing the thoughts of Darwin, he pointed out that what

> prevents us from recognizing the successive changes by which known animals have been diversified … is … that we can never witness these changes. Since we only see the finished work and never see it in course of execution, we are naturally prone to believe that things have always been as we see them rather than that they have gradually developed.[10]

This is an entirely reasonable statement by Lamarck, bearing in mind that variations may take place over an enormous timescale. Finally, anxious not to 'tread on the toes' of God the Creator, Lamarck is quick to affirm that 'doubtless nothing exists but by the will of the Sublime Author of all things',[11] and that an 'assemblage [i.e. a life form] …, is only immutable so long as it pleases her Sublime Author to continue her existence …'.[12]

Chapter 25

Patrick Matthew

On 7 April 1860 a letter appeared in the *Gardeners' Chronicle* which Darwin read, and which must surely have given him an enormous jolt.

> NATURE'S LAW OF SELECTION
>
> TRUSTING to your desire that every man should have his own, I hope you will give place to the following communication.
>
> In your Number of March 3[r]d I observe a long quotation from the *Times,* stating that Mr Darwin 'professes to have discovered the existence and *modus operandi* of the natural law of selection,' that is, 'the power in nature which takes the place of man and performs a selection, *sua* sponte [of its own accord],' in organic life. This discovery recently published as 'the results of 20 years' investigation and reflection' by Mr Darwin turns out to be what I published very fully and brought to apply practically to forestry in my work '[On] Naval Timber and Arboriculture,' published as far back as January 1, 1831, by Adam & Charles Black, Edinburgh, and Longman & Co., London, and reviewed in numerous periodicals, so as to have full publicity in the 'Metropolitan Magazine,' the 'Quarterly Review,' the 'Gardeners' Magazine,' by Loudon [see below], who spoke of it as *the* book, and repeatedly in the 'United Service Magazine [*United Service Journal and Naval and Military Magazine*] for 1831, &c.

The author of the letter, Patrick Matthew, then proceeds to quote extensively from his book.

According to the Scottish zoologist William T. Calman,

> Patrick Matthew was born on 20 October 1790 at Rome, a farm held by his father, John Matthew, on the banks of the Tay near Scone Palace [Perthshire,

> Scotland]. His mother, Agnes Duncan, was related, though in what degree is not known, to the family of Admiral Duncan, the famous ancestor of the present Earl of Camperdown. [British Admiral Adam Duncan (1731–1804), who defeated the Dutch in the Battle of Camperdown (1797).] From her [his mother] he inherited the estate of Gourdiehill before attaining his twentieth year. He was educated at Perth Academy and at the University of Edinburgh, but his stay at the latter cannot have been of long duration, for, on his father's death, he undertook at the age of seventeen, the management of the estate of Gourdiehill, near Errol, Perthshire.[1] This estate he inherited from the Duncan family, in whose possession it had been for more than 300 years. One of his first employments there was the planting of an extensive orchard …

Having inherited Gourdiehill, Matthew commenced the transformation of much of its farmland into orchards devoted to the growing of apples and pears. (The 1831 *Quarterly Journal of Agriculture* contained a seven-page article on the pruning of trees by a Mr Matthew of Gourdie-Hill.)[2] 'In 1817 he married his cousin, Christian Nicol, whose mother Euphemia, was a sister of Agnes Duncan.'[3]

Perhaps on account of his family connections to the Royal Navy, and his own period of life at sea (see below), Matthew became interested in how to grow trees suitable for use in the production of warships. The outcome was that, in 1831, his book *On Naval Timber and Arboriculture* was published. In it he declared that

> It is only on the *Ocean* that *Universal Empire* [i.e. the British] is practicable – only by means of *Navigation* that all the world can be subdued or retained under one dominion.[4]

To return to the year 1860, on 21 April, two weeks after the publication of Matthew's letter in the *Gardeners' Chronicle*, the following letter from Darwin himself, was published:

> I have been much interested by Mr Patrick Matthew's communication in the number of your paper dated April 7th. I freely acknowledge that Mr Matthew has anticipated by many years the explanation which I have offered of the origin of species, under the name of natural selection. I think that no one will feel surprised that neither I, nor apparently any other naturalist, had heard of Mr Matthew's views, considering how briefly they are given, and that they appeared in the appendix to a work on Naval Timber and Arboriculture. I can do no more than offer my apologies to Mr Matthew for my entire ignorance of this publication. If another edition of my work is called for, I will insert a notice to the foregoing effect.[5]

The *Gardeners' Chronicle* published the following statement by Matthew on 12 May, describing how he had reached his conclusions.

> To me the conception of this law of Nature came intuitively as a self-evident fact, almost without an effort of concentrated thought. Mr Darwin here seems to have more merit [i.e. to have been afforded more credit] in the discovery than I have had; to me it did not appear a discovery. He seems to have worked it out by inductive reason, slowly and with due caution to have made his way synthetically from fact to fact onwards; while with me it was by a general glance at the scheme of Nature that I estimated this select production of species as an *à priori* recognizable fact — an axiom requiring only to be pointed out to be admitted by unprejudiced minds of sufficient grasp.[6]

Matthew subsequently complained that an article published on 24 November in the *Saturday Analyst and Leader*, was

> scarcely fair in alluding to Mr Darwin as the parent of the origin of species, seeing that I published the whole [i.e. everything] that Mr. Darwin attempts to prove, more than twenty-nine years ago.[7]

Nonetheless, Darwin kept his promise to Matthew, and when the third edition of *Origin* was published in April 1861, it contained the following statement

> In 1831 Mr Patrick Matthew published his work on *Naval Timber & Arboriculture*, in which he gives precisely the same view on the origin of species as that propounded by Mr Wallace and myself in the *Linnean Journal* and as that enlarged on in the present volume. Unfortunately the view was given by Mr Matthew very briefly in scattered passages in an Appendix to a work on a different subject, so that it remained unnoticed until Mr Matthew himself drew attention to it in the *Gardeners' Chronicle* on April 7th 1860. The differences of Mr Matthew's view from mine are not of much importance … . I am not sure that I understand some passages; but it seems that he attributes much influence to the direct action of the conditions of life [i.e. the environment]. He clearly saw, however, the full force of the principle of natural selection.[8]

Was Matthew correct in stating that his book had been reviewed in 'numerous periodicals'?

The answer is, yes, in the case of three at any rate (although it has not been possible to trace the review in the *Metropolitan Magazine* – a London monthly journal of literature, science, and the fine arts):

a) ***Quarterly Review*** (a literary and political periodical)
Here, the entry, published in the April-July 1833 number under the heading 'On Naval Timber and Arboriculture; with critical notes by Patrick Matthew', was concerned principally with timber rot.

> The author … introduces one of the most important branches of his subject in these terms: we greatly wonder that something efficacious has not been done by our Navy Board in regard to dry rot; and consider that a *rot-prevention-officer wood physician* should be appointed to each vessel of war, from the time the first timber is laid down, to be made accountable if rot to any extent should occur … .

The reviewer also admits that Matthew's book 'on the whole, is not a bad one …'.[9]

b) ***Gardeners' Magazine*** *– and Register of Rural and Domestic Improvement* (a periodical devoted solely to horticulture, for the benefit of gardeners throughout the country and founded in 1826 by Scotsman John Claudius Loudon, garden designer and author).[10] Extracts from its 1832 review of Matthew's book are quoted below.[11]

> The author introductorily maintains that the best interests of Britain consist in the extension of her dominion on the ocean; and that, as a means to this end, naval architecture is a subject of primary importance; and, by consequence, the culture and production of naval timber is also very important. He explains, by description and by figures, the forms and qualities of planks and timbers most in request in the construction of ships; and then describes those means of cultivating trees, which he considers most effectively conducive to the production of these required planks and timbers.
>
> 'The British forest trees suited for naval purposes,' enumerated by the author, are, oak, Spanish chestnut, beech, Scotch elm, English elm, red-wood willow (*Sâlix frágilis*), red-wood pine, and white larch. On each of these he presents a series of remarks regarding the relative merits of their timber ….
>
> Two hundred and twenty-two pages are occupied by 'Notices of authors relative to timber' [i.e. critical reviews of the works of other authors].

Finally, a brief mention is made of the 'Appendix' in which Matthew claimed to have proposed his 'theory of natural selection'.

> An appendix of 29 pages concludes the book, and receives some parenthetical evolutions of certain extraneous points which the author struck upon in prosecuting the thesis of his book. This may be truly termed, in a double sense,

> an extraordinary part of the book. One of the subjects discussed in this appendix is the puzzling one, of the origin of species and varieties; and if the author has hereon originated no original views (and of this we are far from certain), he has certainly exhibited his own in an original manner.[12]

c) ***United Service Journal and Naval and Military Magazine***
This review ran to approximately 13,000 words and appeared in two parts – in the 1831 Part II and the 1831 Part III numbers of the magazine respectively. The reviewer was generally appreciative of Matthew's book, stating that

> The British Navy has such urgent claims upon the vigilance of every patriot, as the bulwark of his independence and happiness, that any effort for supporting and improving its strength, lustre, and dignity, must meet with unqualified attention.[13]

No mention was made of the book's 'Appendix'.

Finally, it should be noted that the *Edinburgh Literary Journal* also reviewed Matthew's book in its number dated 2 July 1831; a fact which, not surprisingly, he failed to mention, as the review was scarcely complimentary.

> This is a publication of as great promise, and as paltry performance, as ever came under our critical inspection. From its title … it will probably attract readers; but the intelligent among them will suffer considerable disappointment in the perusal, as we must say that there are not ten pages of really *new* matter in the volume … .

Can Darwin be blamed for not having been aware of the existence of Matthew's book *On Naval Timber and Arboriculture*, and of the significance of its contents?

i. Darwin was abroad, voyaging on HMS *Beagle* from the time the ship set sail on 27 December 1831 until her return on 2 October 1836. He was, therefore, thousands of miles from home, both when Matthew's *On Naval Timber & Arboriculture* was published in January 1831, and also when the book was reviewed in 1831/2/3. He may, therefore, be forgiven for being unaware of the book's existence during this period.

ii. In fairness to Darwin, the book's title *On Naval Timber and Arboriculture*, gave no indication that, as Matthew alleged, its 'Appendix' contained a ground-breaking new theory about 'natural selection'.

iii. Henslow, at the time in question, was a regular contributor to *Gardeners' Magazine*, and a paper by him entitled 'Henslow's examination of a Hybrid Digitalis',[14] appeared in *the very same volume in which the review of Matthew's book is to be found.* And in Volume XIII (1837) Henslow is mentioned on no fewer than eight separate occasions. As for Hooker, it is equally difficult to imagine that anything of any significance which appeared in *Gardeners' Magazine* would have escaped his notice. However, if they were aware of Matthew's book (and of its reviews), none of Darwin's colleagues drew his attention to it, even after his return to England in late 1836.

iv. It was not obvious to any of the reviewers of Matthew's book that it contained a new theory about natural selection. (However, as already mentioned, the review in *Gardeners' Magazine* did refer to the fact that 'the origin of species and varieties' had been discussed by Matthew in his Appendix.)

Was Darwin telling the truth when he claimed only to have become aware of Matthew's book in April 1860 – i.e. twenty-nine years after its publication?

a. The fact that a copy of *On Naval Timber & Arboriculture* is to be found in Darwin's personal library with an autograph signature 'C. Darwin Apr. 13th 1860' inside the front cover of the volume, corroborates Darwin's assertion that he was telling the truth in this matter.[15]

b. Volumes 2–13 (1827–37) of *Gardeners' Magazine* (i.e. including the 1832 number which contained the review of Matthew's book) were also to be found in Darwin's personal library. However, inside the front board of volume 7 (for 1831), at the top left-hand corner, are to be found the initials of Robert Waring Darwin, Charles's father. Clearly, therefore, these volumes (which include those from 1831–36 when Darwin was at sea on HMS *Beagle*) were acquired by Robert for his library at 'The Mount' in Shrewsbury,[16] and Darwin presumably acquired them only after his father's death in November 1838.

Finally, and crucially, was Darwin correct in believing that Matthew had 'got there first', in respect of his great theory? This will be discussed shortly.

* * *

***On Naval Timber and Arboriculture* by Patrick Matthew**

This is a substantial volume of approximately 90,000 words in which the author does not confine himself solely to the subject of timber, but devotes six pages to the subject of 'universal free trade'; to the 'absolute necessity of abolishing every monopoly and restriction of trade in Britain', and speaks of the 'insane duty on the importation of naval timber and hemp'.[17] He also includes a vociferous rant against 'the law of entail [which requires that property and wealth remains within a family]', which, he said, was 'an outrage'.[18]

When Darwin described some passages of Matthew's work *On Naval Timber* as being 'rather obscure', this is not surprising, as the latter's writings often lack clarity of meaning. (Heretical as it may seem to say so, Darwin himself was guilty, on occasion, of the same fault!) However, when Matthew speaks of the 'infinite variety existing in what is called species',[19] which applies to 'vegetables as well as animals …',[20] Darwin would surely have agreed.

Matthew's so-called law of self-adaptation

In respect of fossils, Matthew remarks how

> Geologists discover … an almost complete difference to exist between the species or stamp of life on one epoch from that of every other. We are therefore led to admit, either of a repeated miraculous creation; or of a power of change, under a change of circumstances, to belong to living organised matter … .[21]
>
> Is the inference then unphilosophic, that living things which are proved to have a circumstance-suiting power – a very slight change of circumstance by culture inducing a corresponding change of character – may have gradually accommodated themselves to the variations of the elements containing them, and, without new creation, have presented the diverging changeable phenomena of past and present organized existence?[22]

Matthew speaks of animals and plants 'in the course of time, moulding and accommodating their being anew to the change of circumstances …',[23] and of 'the self-regulating adaptive disposition of organized life …'.[24] In other words, it was Matthew's view that, rather than having been spontaneously created, animals and plants had, by virtue of their own inherent ability, adapted themselves in order to suit the environment in which they found themselves, and had diverged in character accordingly.

Finally, so confident is Matthew of his theory that he goes so far as to describe it as 'this self-adaptive law'.[25]

Conclusion

When, in his book, Matthew made reference to 'A principle of selection existing in nature of the strongest varieties for reproduction',[26] he was in no way proposing a theory of natural selection in the Darwinian sense (and Darwin *was incorrect in believing him to have done so*). Instead, he was referring to an alleged (but unproven) innate ability of certain organisms to self-adapt to their particular environment.

Darwin himself believed that changes in the environment had a part to play in inducing variations to occur in organisms, and modern research suggests that yes, this might well be the case, as will shortly be seen.

Sequel

Darwin wrote to Matthew on 13 June 1862 as follows:

> Dear Sir – I presume that I have the pleasure of addressing the Author of the work on Naval Architecture and the first enunciator of the theory of Natural Selection. Few things would give me greater pleasure than to see you; but my health is feeble and I have at present a son ill and can receive no one here, nor leave home at present.
>
> I wish to come up to London as soon as I can; if, therefore, you are going to stay for more than a week, would you be so kind as to let me hear, and if able to come up to London, I would endeavour to arrange an interview with you, which [would] afford me high satisfaction. With much respect I remain, dear sir, yours very faithfully Ch. Darwin.[27]

Two years later, yet another offer to meet was made by Matthew, which Darwin was obliged to decline on the grounds of his own ill health.

Chapter 26

William Charles Wells

In October 1865 Darwin wrote to Hooker in respect of a learned paper, published no less than forty-seven years previously, which had just been brought to his notice.

> In 1813 Dr W. C. Wells read before the Royal Society 'An Account of a White Female, part of whose Skin resembles that of a Negro'; but his paper was not published until his famous 'Two Essays upon Dew and Single Vision' appeared in 1818 [i.e. the year after Wells's death]. In this paper[1], he distinctly recognizes the principle of natural selection, and this is the first recognition which has been indicated; but he applies it only to the races of man, and to certain characters [i.e. characteristics] alone.[2]
>
> I am indebted to Mr Rowley, of the United States, for having called my attention, through Mr Brace, to the above passage in Dr Wells's work.[3]

(Charles Loring Brace was a US anthropologist from the city of Hudson, New York, and Robert S. Rowley was his friend and neighbour.) Darwin therefore concluded that

> poor old Patrick Matthew is not the first, and he cannot, or ought not, any longer to put on his title-pages 'Discoverer of the principle of Natural Selection'![4]

Wells was born in Charles Town (Charleston), South Carolina in 1757; his parents, Robert, a printer and bookseller and Mary, having emigrated to the USA from Scotland. However, Wells returned to Scotland to attend school. From 1775–78 he attended Edinburgh University where he studied medicine (and subsequently graduated as MD in 1780).[5] His career was to be a distinguished one, culminating in his election to the post of Physician to London's St Thomas's Hospital, and to the Royal Societies of both London and Edinburgh.

The full title of Wells's paper, to which Darwin refers, is 'An account of a female of the white race of mankind, part of whose skin resembles that of a negro, with some observations on the cause of the differences in colour and form between the white and negro races of man'. Its subject was Hannah West of Sussex, who had been admitted to St Thomas's Hospital 'on account of some bodily ailment', and who had 'blackness on part of her skin, which was observed at her birth'. This latter fact led Wells into a discussion as to the possible role of skin colour in relation to survival in hot or cold climates. First, however, Wells stated that

> those who attend to the improvement of domestic animals, when they find individuals possessing, in a greater degree than common, the qualities they desire, couple a male and female of these together, then take the best of their offspring as a new stock, and in this way proceed, till they approach as near the point of view [objective], as the nature of things will permit.

Now comes the crucial sentence.

> But, what is here done by art, seems to be done, with equal efficacy though more slowly, by nature, in the formation of varieties of mankind, fitted for the country which they inhabit.

Wells now gives a practical example of his theory.

> Of the accidental varieties of man, which would occur among the first few and scattered inhabitants of the middle regions of Africa, some one would be better fitted than the others to bear the diseases of the country.
>
> This race would consequently multiply, while the others would decrease, not only from their inability to sustain the attacks of disease, but from their incapacity of contending with their more vigorous neighbours. The colour of this vigorous race I take for granted, from what has been already said, would be dark. But the same disposition to form varieties still existing, a darker and a darker race would in the course of time occur, and as the darkest would be the best fitted for the climate, this would at length become the most prevalent, if not the only race, in the particular country in which it had originated.
>
> In like manner, that part of the original stock of the human race, which proceeded to the colder regions of the earth, would in process of time become white, if they were not originally so, from persons of this colour being better fitted to resist the diseases of such climates, than others of a dark skin.[6]

This demonstrates, conclusively, that Wells had indeed grasped both the phenomenon of variation, and most importantly, the principle of natural selection.

'The negro race is better fitted to resist the attacks of the diseases of hot climates than the white,' Wells asserted, but he was not naïve enough to suppose that the 'different susceptibility' of the black and white races of man to disease was purely the result of 'their difference in colour'. This was due, he said, 'to some difference in them … [as] yet too subtle to be discovered.'

Darwin was, therefore, correct in regarding Wells as being the first known person to have recognized the principle of natural selection, albeit in respect of human beings only.

.

Chapter 27

Darwin's Chronic Illness: Dr James M. Gully

For all of his adult life Darwin suffered from ill health. Various suggestions have been made as to the reason for this, and it has even been implied that his symptoms were of psychoneurotic origin (see below). So is it possible to discover the truth of the matter?

In the absence of any surviving medical records it is necessary to rely on the words of Darwin himself, over the years, for a description of his medical condition. In his correspondence and in his diary, he described his symptoms factually, in the manner of the trained scientist, as might be expected. He did not grumble, per se, or express self-pity, but rather bore his physical discomforts stoically. However, this did not prevent him from expressing exasperation at having to convalesce when his work beckoned, and anxiety lest he become a burden to his family. And despite his own sufferings, he was always concerned for the health and well-being of others.

During the two months spent at Plymouth prior to the departure of HMS the *Beagle* on 27 December 1831, Darwin declared that he was

> troubled with palpitation and pain about the heart, and like many a young ignorant man, especially one with a smattering of medical knowledge, was convinced that I had heart disease.[1]

The following year in early summer, when *Beagle* with Darwin aboard was in the vicinity of the Brazilian city of Rio de Janeiro, Captain FitzRoy stated,

> Upon making … inquiry respecting those streams which run into the great basin of Rio de Janeiro, I found that the Macacu was notorious among the natives as being often the site of pestilential malaria, fatal even to themselves … .[2]

This indicates just how dangerous the region could be for the crew of HMS *Beagle* (Darwin included), some of whom paid with their lives.

On 14 October 1837, a year after the return of HMS *Beagle*, Darwin remarked:

> Of late, anything which flurries me completely knocks me up afterwards and brings on a b[ad] palpitation of the heart.[3]

Between January 1839, which was the month before he married Emma, and September 1842,

> I did less scientific work, though I worked as hard as I possibly could, than during any other equal length of time in my life. This was owing to frequently occurring unwellness, and to one long and serious illness.[4]

In the summer of 1842 Darwin undertook a solo expedition to North Wales with the object of

> observing the effects of the old glaciers which formerly filled all the larger valleys … [This] was the last time I was ever strong enough to climb mountains or to take long walks such as are necessary for geological work.[5]

Why should this be? How could a man of only thirty-three years of age, who had showed such stamina and resilience during the *Beagle* voyage, have deteriorated to such an extent?

Having relocated to Down House on 14 September, Darwin declared that

> During the first part of our residence we went a little into society, and received a few friends here; but my health almost always suffered from the excitement, violent shivering and vomiting attacks being thus brought on.[6]

Sadly, therefore, the move to the country had failed to achieve the desired restorative effect.

On several occasions throughout his life Darwin refers to problems with his hands. For example, to Emma in October 1843, he speaks of a 'dreadful numbness in my finger ends …'.[7]

In August 1844,

> My health during the last three years has been exceedingly weak, so that I am able to work only two or three hours in the 24 … .[8]

In early February 1845,

> I have given up acids & gone to puddings again … My stomach is baddish again this morning … .[9]

And again on 31 March,

> I believe I have not had one whole day or rather night, without my stomach having been greatly disordered, during the last three years, & most days great prostration of strength … .[10]

While in early September,

> I have taken my Bismuth [an ingredient of medicines used to treat gastro-intestinal diseases] regularly, I think it has not done me quite so much good, as before … .[11]

To Hooker, in November, he wrote,

> You ask about my health; I have been unusually well for a week past, owing, I believe, to what sounds a great piece of quackery, viz. twice a day passing a galvanic stream through my insides through a small-plate battery for half an hour [a practice commonly known as 'galvanization']. I think it certainly has relieved some of my distressing symptoms.[12]

Darwin wrote to Emma, who was staying with the children at Tenby on the South Wales coast on 25 June 1846 to say,

> I have been stomachy & sick again, but not very uncomfortable; I will take the blue-pill [presumably bismuth] again.[13]

Then, in early July,

> I have had a good deal more sickness than usual.[14]

And on 3 September,

> I have of late been slaving extra hard, to the great discomfiture of [my] wretched digestive organs… .[15]

Hooker received an apology from Darwin on 7 April 1847.

> I should have written before now, had I not been almost continually unwell, & at present I am suffering from four boils & swellings, one of which hardly allows me the use of my right arm & has stopped all my work & dampened all my spirits.[16]

To the Swiss geologist Bernhard Studer, Darwin wrote on 4 July,

> Shortly after my return from my long voyage, I had a tedious & severe illness, & I have never since recovered my strength & suppose I never shall … . I appear quite well, but from being a strong man, I am become incapable of any continued muscular exertion; or indeed of much exertion of mind, for even conversation, if it excites me, tires me in a very short time, so that I am compelled to live a most retired life.[17]

In October, Darwin told Emma that he was still being plagued by boils.[18]

In late April/early May 1848, Darwin described to Emma a particularly severe bout of illness.

> My attack was very sudden: it came on with fiery spokes & dark clouds before my eyes [presumably abnormal visual images]; then sharpish shivery & rather bad [or] not very bad sickness. I got up yesterday about 2, & about 7 I felt rather faint & had a slight shaking fit & [a] little vomiting & then slept too heavily; so today I am languid & stomach bad … .[19]

In November, wrote Darwin,

> So much was I out of health that when my dear father died on November 13th … I was unable to attend his funeral or to act as one of his executors.[20]

This speaks volumes for the severe degree of Darwin's incapacity, bearing in mind his devotion to his family.

Darwin later told Hooker that,

> All this winter I have been bad enough, with dreadful vomiting every week, & my nervous system began to be affected, so that my hands tremble & head was often swimming.[21]

Darwin's vomiting attacks appear to have occurred without warning, giving him too little time to reach the lavatory. If such an attack occurred when he was at work in his study, he was therefore obliged to pull back the screen and vomit into the washbasin, which had been installed within a cubby hole in that room.

On 24 February 1849 Darwin told Professor Richard Owen that he intended to visit Malvern Wells, Worcestershire, to see 'whether there is any truth in [Dr] Gully & the water cure: regular Doctors cannot check my incessant vomiting'.[22]

Dr James Manby Gully and his Water-cure
Gully was an advocate of clairvoyance, homeopathy, and hydropathy. In his book *The Water-cure in Chronic Diseases*, published in 1880, he gave details of what the *Water-cure* entailed, and what its beneficial effects were likely to be.

Hot and warm fomentations
A warm flannel is

> placed over the part to be fomented … . The most ordinary place for applying this process is on the belly and especially the portion of it between the bottom of the breastbone and the navel, and across [and] far back on both sides … . In this manner you include … *the stomach* in all its length … .

Packing in damp towels and sheets
A towel is placed on 'the front of the trunk' and subsequently on the back of the trunk, and warmed to an

> appropriate temperature. … it is expedient to wring the sheet out of warm water, and have it applied around the body at a temperature of about 70° or 75°.

The sitz bath [in German *sitzbad* – a bath in which a person sits in water up to the hips – from the words *sitzen* (sit) and *bad* (bath)]
The patient is immersed in cold water 'for a time varying from five to fifteen minutes', it being necessary to repeat this process 'as often as six or seven times in the 24-hours'.

The abdominal compress

> The object of the compress is to produce and maintain over the abdominal viscera an amount of moist warmth which shall act as a counteracting and soothing agent to the irritation which is fixed in those viscera. [Such a compress] should be worn only for two or three hours at a time, or should be frequently refreshed from cold water.

The dripping or rubbing sheet
The patient is enveloped in a cold or tepid sheet which serves the same function as having a tepid bath.

The shallow bath

> Seated in eight or ten inches of water, with the legs extended in it, the patient is sponged and splashed with the water of the bath ….

The douche

> The stimulus afforded by the repeated changes of [cold] water is very much greater [than with the shallow bath], for the water is pouring incessantly upon the body and therefore is incessantly changing. By this, too, a great amount of heat is withdrawn from the surface.

The sweating process

> In this process, the excitation of the whole nervous system and of the circulation is produced by accumulated *heat* applied to the surface; and although in this it differs from the douche, which excites by the incessant application of *cold*, the result upon the functions in question is pretty nearly the same.

There were also 'foot and hand baths, and minor ablutions and frictions' to be undertaken.

Water drinking

Gully recommended the drinking of 'from three to six tumblers daily' of water, 'and that should be taken in very divided quantities …'.

> In some of the worst instances of nervous indigestion, the great centre of the nutritive nerves is so exquisitely sensitive, that the shock of even half a tumbler of cold water upon the stomach is transmitted to the brain and there causes giddiness, confusion, nervous aching &c … . On this account I have now and then raised the water [temperature] to 55° or 58°, when the drinking of it was indispensable, until the nerves of the stomach became more able to bear the natural temperature … .

Diet

> Experience gives me no room to doubt that by appropriate regulation of the diet to each case, restoration is secured in much less time, and at much less of that constitutional term, tumult, which harsh practice rouses.

Under the heading 'Diet Table for Patients under Water Treatment: Things Permitted' were included a variety of soups, including 'plain beef, mutton or chicken broth … or with the additional of carrots, young peas, cauliflower, rice' etc.; a variety of fish dishes; 'Meat and animal products, including beef, mutton, *lean* pork, veal, venison' etc., and various 'vegetables and roots'.

Also permitted were

> milk, but without cheese; Sweets and fruits, eaten with butter or sugar [and] drinks [including] water, toast and water, barley or rice water and sometimes milk, milk and water [and] weak black tea, almost cold.

However, said Gully,

> I have to forbid some patients the use of animal food [presumably meat and dairy products] three or four days in the week, and others for a week together: to some I forbid all puddings, even farinaceous [farina being flour or meal derived from cereal, nuts, or starchy roots], after meat: to others all vegetable matter but bread, &c. All this is subject for [i.e. to] weekly or even daily change; and it is impossible to lay down rules applicable to all cases.

Clothing

> It is desirable to avoid all clothing which shall have for view [i.e. as its purpose] to keep the body in a state of artificial heat.

Habits of life

> The sufferer from chronic disease … must learn to rise early and to walk or work so as to gain appetite: when this appetite is acquired he must eat, whether the hour be fashionable or not: and he must go to bed early from the motive of fatigue alone.[23]

From Gully's own description, the *Water-cure* was clearly not for the faint-hearted. 'Torture' might be deemed too strong a word, but it was certainly a severe ordeal to be undergone. For Darwin, what small benefit was obtained proved to be of a purely transient nature, which indicated that even when his symptoms were temporarily ameliorated, the underlying disease process – whatever its nature might be – was continuing unabated.

* * *

Darwin told Hooker on 28 March 1849 that, on his return to Down House from Malvern, he had persisted with Dr Gully's treatment which involved being

> heated by spirit lamp till I *stream* with perspiration, & am then suddenly rubbed violently with towels dripping with cold water: have two cold feet-baths, & wear a wet compress all day on my stomach.[24]

In that same year, Darwin

> erected a waist-high tub within a specially constructed wooden shelter near the [water] well… . Here, in pursuit of a cure for his persistent stomach problems, he showered with ice-cold water from the cistern above, and was then scrubbed from head to foot by the faithful [butler] Parslow.[25]

Such were the lengths to which Darwin was prepared to go in order to improve his health. Alas, it was largely to no avail.

Chapter 28

Darwin's Continuing Ill-Health: Possible Causes

Despite his efforts, and those of the physicians who attended him, Darwin continued to suffer the frustrating and debilitating effects of his mysterious illness.

In March 1851 Darwin confessed that

> in old days my greatest pleasure was the conversation of scientific men, but I find by dear-bought experience that I cannot visit anywhere, as the excitement invariably does me harm for days afterwards.[1]

And in October,

> All excitement & fatigue brings on such dreadful flatulence; that in fact I can go nowhere. The other day I went to London & back, & the fatigue, though so trifling [brought] on my bad form of vomiting.[2]

Darwin suffered from chronic flatulence – this being defined as an accumulation of gas in the alimentary tract.[3] This fact may have been an additional reason – other than his lassitude and malaise – why he shunned company and public meetings, i.e. out of potential embarrassment to himself.

Two years later, in July 1853, Darwin told his cousin William D. Fox, 'my bug-bear is hereditary weakness'.[4] Darwin was aware that there had been inbreeding in the family, and was also aware of the potential deleterious effects which this might have. Therefore, in the absence of any other specific suggestion by him as to the nature of this hereditary illness, it is assumed that it was to the mysterious illness described above to which he was referring. The subject will be discussed in more detail later.

On 13 March the following year, in his so-called, 'Diary of Health',[5] Darwin recorded that he felt 'Very Poorly much vomiting Bad Boil'.[6] He kept this diary from January 1849 to 30 January 1855, during which six-year period he reported frequent 'fits of flatulence'.

The following day he told Hooker that

> I am very unwell & must write briefly, for I am & have been suffering from an immense Boil or almost [an] abscess.[7]

In his autobiography Darwin said of *A monograph on the fossil Balanidae and Verrucidae of Great Britain*, published in 1855 that,

> Although I was employed during eight years [i.e. from 1846 until 1854] on this work, yet I record in my diary that about two years out of this time was lost by illness.[8]

In that year of 1855 Darwin describes outbreaks of eczema which affected his face. This led Emma to suggest that he grew a beard in order to disguise it – advice that he accepted.[9] Because no detailed description exists of the nature of Darwin's eczema, Dr Ralph Colp (see below) agrees with the diagnosis of US dermatologist Gordon Sauer that this had an allergic basis that was not related to the other symptoms from which Darwin suffered.[10]

In November Darwin wrote to Hooker to say that he was hoping to attend a meeting of the Council of the Royal Society but would not be able to do so if he happened to have a headache.[11] (It should be noted that this was one of the few occasions when Darwin complained of a headache.)

Darwin made a request to Fox, in October 1856, that he (Darwin) be '*most* kindly' remembered to Dr Gully, and to tell him that 'never (or almost never) the vomiting returns, but that I am a good way from being a strong man'.[12] Alas, the benefit was to be short lived.

In April 1857 he wrote that

> My health has been very poor of late & I am going in a week's time for a fortnight of hydrotherapy & rest. This was to be at [physician and hydrotherapist] Dr Edward Wickstead Lane's hydropathic establishment at Moor Park, Farnham, Surrey.[13]

Sadly, whatever benefit Darwin may have derived was again short lived, for a month later, he wrote

> It is very disheartening for me, that all the wonderful good which Moor Park did me at the time, has gone away like a flash of [lightning] now that I am at work again.[14]

And in early June he declared that

> It is most provoking that a cold on leaving Moor Park suddenly turned into my old vomiting, & I have been *almost* as bad since my return home as before … .[15]

In April 1858,

> My health has been lately wretched … .[16]

Later that month Darwin was to be found once again at Moor Park 'for a fortnight's hydropathy as my stomach had got from steady work into a horrid state'.[17] (The terms 'hydropathy' and 'hydrotherapy' are synonymous, meaning the treatment of illness through the use of water, either internally or through external means such as steam baths.[18] Exercises in a pool may also be part of the treatment.) On 28 April a seemingly improved Darwin wrote, in relaxed and cheerful mood, to Emma from Moor Park to say that

> The weather is quite delicious. Yesterday after writing to you I strolled a little beyond the glade for an hour & half & enjoyed myself … . At last I fell fast asleep on the grass & awoke with a chorus of birds singing around me, & squirrels running up the trees & some Woodpeckers laughing, & it was as pleasant a rural scene as ever I saw, & I did not care one penny how any of the beasts or birds had been formed.[19]

For once, he was relaxing and enjoying nature, rather than making a study of it.

On 8 May, by which time he had returned to Down, Darwin told Fox

> I got splendidly well the last few days at Moor Park, & walked one day two miles out & back; & I am now hard at work again as usual.[20]

But a month later,

> I am confined to sofa with Boil … .[21]

In the new year of 1859 a despairing Darwin declared that

> My health keeps very poor & I never know 24 hours comfort. I force myself to try & bear this as incurable misfortune.[22]

From Moor Park, to which he had returned despite the lack of progress, Darwin wrote to John Phillips, Deputy Reader in Geology at Oxford University, on 8 February to say that he was too ill to receive the award of the Wollaston Medal of the Geological Society, but would 'ask Lyell to receive the medal for me …'.[23] Four days later he reported being

> extra bad of late, with the old severe vomiting rather often & much distressing swimming in the head.[24]

In May Darwin stated that his health had 'quite failed'. He therefore left Down for Moor Park for another week of hydrotherapy.[25] From there, on 26 May, he wrote to Hooker to say, 'I had bad vomiting before starting [the treatment] & great frustration of mind & body….'[26]

For the period from 2 October until 7 December, Darwin was resident at the Wells House Hydropathic Establishment, Ilkley, Yorkshire, for a course of homeopathy under the care of its proprietor physician, Dr Edmund Smith.[27] On 16 November he told Fox that he had suffered 'a frightful succession of Boils – 4 or 5 at once'.[28] And on Christmas Day,

> I was hardly able from lameness, Boils &c to give Water-cure [hydrotherapy] a fair trial this time … .[29]

He told Hooker on 22 January 1860 that 'I mean to come to London on Tuesday evening for the vain purpose of consulting a new Doctor for my stomach … .' The doctor in question, according to Darwin's account book, was a Dr Edward Headland, 'the leading general physician in London'.[30] However, Darwin's tone implies that he had no confidence in any cure being found.

Darwin wrote to his son William on 22 March to say that

> I am trying under a Mr [Dr] Headland a course of nitro-muriatic acid [an archaic term for hydrochloric acid], eating no sweet things & drinking some wine: but it has done nothing for me as yet & I shall go to my grave, I suppose, grumbling & groaning with daily, almost hourly, discomfort.[31]

By 25 June he was in despair.

> My stomach has utterly broken down & I am forced to go on Thursday for a little water-cure, to Dr Lanes Sudbrook Park, Richmond, Surrey [where Dr Edward Wicksted Lane had opened a new hydropathic establishment], where I shall stay a week … .[32]

On 23 April 1861 Darwin told Hooker,

> I spoke a few minutes at [Linnean Society] on Thursday & though extra well, it brought on 24 hours vomiting.[33]

Darwin declared, in late August, that a recent holiday in Torquay, Devonshire with the family had resulted in 'some benefit to my health & much good to my daughter's [Henrietta]'.[34]

Darwin wrote to Hooker in mid-September 1862 from Cliff Cottage – a modest, thatched abode in Bournemouth where he was holidaying, to say:

> I have been thinking of [visiting] Cambridge, for a few days, & your going [there] is an immense temptation, but I very much fear I shall not be strong enough; I have had [a] headach[e] half [of] every day, with my stomach intolerably bad.[35]

In January 1863 Darwin told Hugh Falconer that his eczema had 'taken off the epidermis [outer layer of the skin] a dozen times clean off …'.[36] later, on 23 March, he told Fox, 'I am having an attack of Eczema on my face … .' He describes 'having had [the] first attack of this horrid & blessed eczema' at Ilkley in Yorkshire (which he had visited in autumn 1859).[37] Three days later he informed Hooker that he had 'consulted the great Mr Startin' for his eczema – James Startin being a leading skin specialist in London.[38]

On 3 November, and again in the following month, William Brinton, Physician to St Thomas's Hospital, London and a specialist in stomach disorders, attended Darwin at Down House.[39] September found him once again at Malvern Wells, this time receiving treatment from Dr Gully's associate, physician and surgeon James Smith Ayerst.[40]

Wrote Darwin in late January 1864, 'I shd suppose few human beings had vomited so often during the last 5 months.'[41] And a month later, 'The vomiting is not now daily … . My head hardly now troubles me, except singing in ears – It is now six months since I have done a stroke of work … .'[42] That spring Darwin was so indisposed, health-wise, that he was obliged to dictate his letters to Emma instead of writing them himself.

In April Darwin consulted Dr William Jenner, Physician Extraordinary to Queen Victoria and Physician to University College Hospital, London who prescribed 'small & very frequent doses of Chalk, Magnesia & Carbonate of ammonia', but with only limited success.[43] That December Darwin wrote:

> I have suffered from almost incessant vomiting for nine months, & that this has so weakened my brain, that any excitement brings on whizzing & fainting feelings, when I cannot speak; & much of this makes me for days afterwards very unwell … .[44]

For a workaholic like Darwin to be denied the opportunity to pursue his studies, and instead to be confined to his bed or couch, or incarcerated in one or other hydropathic establishment, must have been unutterably frustrating and vexatious. And, to make matters worse, some of the finest physicians in the land could offer no convincing diagnosis, let alone a cure.

Darwin was visited at Down House on 20 May 1865 by yet another in a long line of medical experts, this time physician, author, and publisher Dr John Chapman, who, with a view to curing his persistent vomiting, recommended the application of ice (contained in a specially designed ice bag) to the spinal area.[45] However, on 7 June, Darwin informed Chapman that 'the ice does not stop either [the] flatulence or sickness … '.[46]

By far the most comprehensive account of his state of health was written by Darwin himself on the very day of Dr Chapman's visit. It reads as follows:

> For [the past] 25 years extreme spasmodic daily & nightly flatulence: occasional vomiting; on two occasions prolonged during months. Extreme secretion of saliva with flatulence, Vomiting preceded by shivering, hysterical crying dying sensations or half-faint. & copious very [pallid] urine. Now vomiting & every paroxys[m] of flatulence preceded by singing of ears, rocking, treading on air & vision. focus & black dots— All fatigues, specially reading, brings on these Head symptoms ?? [the question marks indicating, presumably, that Darwin did not know what was causing these symptoms] nervousness when E. [Emma] leaves me.
>
> (What I vomit [is] intensely acid, slimy (sometimes bitter) corrodes teeth.)
>
> Doctors puzzled, say suppressed gout [a vague and ill-defined condition] Family gouty. No organic mischief [according to Drs] Jenner & Brinton [whom he had previously consulted].
>
> Tongue crimson in morning ulcerated— stomach constricted dragging. Feet coldish.— Pulse 58 to 62—or slower & like thread. Appetite good — not thin. Evacuation [of bowels] regular & good. Urine scanty (because do not drink) often much pinkish sediment when cold— seldom headache or nausea.
>
> Cannot walk above ½ mile — always tired — conversation or excitement tires me most.
>
> Heavy sleep— bad day.
>
> Eczema—(now constant) lumbago— fundament [buttocks]— rash.
>
> Always been temperate— now wine comforts me much— could not take any formerly. Physic no good— Chalk & Magnesia.— Water-cure & Douche. Last time at Malvern could not stand it –
>
> I fancy that when much sickness my stomach is cold— at least water is very little warmed.

I feel nearly sure that the air is generated somewhere lower down than the stomach & as soon as it regurgitates into the stomach the discomfort comes on.
Does not throw up the food.
Instruction— How soon any effect? How long continue treatment?[47]

Although this account by Darwin is, at times, difficult to interpret, he describes his commonest symptoms as being flatulence, vomiting, fatigue, visual and auditory disturbances, and anxiety. However, his appetite remains good, and he has not lost weight ('not thin').

In July, and again in August, Darwin consulted physician and chemist Henry Bence Jones, who recommended a diet which consisted of 'almost exclusively bread & meat'.[48] On 7 September he informed the surveyor, civil engineer, and architectural draftsman Edward Cresy, who had advised him with regard to alterations to be made to Down House, 'I have been confined to my bed-room with almost constant sickness for the last four months, and have seen nobody and done nothing.'[49] On the 27th he told Hooker that Bence Jones's diet 'has done me good' and that 'my vomiting is immensely reduced'.[50] However, on 25/26 October, he told Fox that Bence Jones 'has stopped my vomiting by a scanty diet of toast & meat, yet I cannot recover my strength.'[51]

In early January 1866 he wrote that

I have not yet much taste for common meat, but eat a little game or fowl twice a day & eggs, omelet or macaroni or cheese at the other meals & these I think suit me best.[52]

In February 1868,

I am at work again, as hard as I am able. It is really a great evil, that from habit I have no pleasure in hardly anything except natural history, for nothing else makes me forget my ever recurrent uncomfortable sensations.[53]

A year later, in February 1869, he wrote, 'My health continues the same; I am always what is called 'a poor Devil'; but I am able to work for two or three hours daily.'[54] and in April, 'I am not very well owing to a fall from my horse ….'[55] Darwin was evidently still able to ride, despite his ill health.

To Hooker, on 22 June, he wrote from Barmouth in North Wales, where he was taking a prolonged holiday with the family:

I have been as yet in a very poor way: it seems as soon as the stimulus of mental work stops, my whole strength gives way: as yet I have hardly crawled

> half a mile from the House, & then been fearfully fatigued. It is enough to make one wish oneself quiet in a comfortable tomb.[56]

Darwin informed Dr Bence Jones, in early August 1870, that his stomach was 'diabolical'.[57]

In late January 1871 he said, 'I was so ill yesterday I hardly knew what I wrote',[58] while, on 15 March, he wrote to Patrick Matthew, 'My health keeps very indifferent & every exertion fatigues me, so that I doubt whether I shall be good for much more [i.e. work]'.[59] And on 6 November,

> My health is very weak: I *never* pass 24 hours without many hours of discomfort, when I can do nothing whatever. I have thus also lost two whole consecutive months this summer.[60]

The evidence clearly points overwhelmingly to the fact that there was something chronically amiss with Darwin's health. But what?

Possible causes of Darwin's illness

1. Cyclic Vomiting Syndrome

In an article entitled 'Darwin's Illness Revisited', published in December 2009, John A. Hayman, Associate Professor, Department of Anatomy and Developmental Biology, Monash University, Melbourne, Australia, stated categorically

> Darwin's symptoms are those of cyclical vomiting syndrome. Although this is primarily a disease of children it may persist into adulthood or may appear for the first time in adulthood.
>
> People with cyclical vomiting syndrome experience abdominal, circulatory, and cerebral symptoms, including headaches and anxiety. Symptoms overlap with those of classic and abdominal migraine, except for a lack of aura [white or coloured light surrounding the object in view].
>
> Affected people may experience some or all of these symptoms, with each individual having similar symptoms with each episode. Over time, however, progression or change may occur in the most prominent feature, and episodes may coalesce.
>
> Episodes of illness may be divided into three phases – prodromal [the period between the appearance of initial symptoms and the full development of clinical symptoms and signs], emetic, and recovery – often with definite triggering events. Symptoms in the prodromal phase include fatigue, palpitations, and sweating. The emetic phase may consist of continuous nausea, with vomiting

> two to 20 times an hour. In most people this is associated with severe abdominal pain. Episodes may last one or two days or up to a week.[61]

There are several reasons why this theory does not fit with the facts. For example, Darwin seldom experienced headaches; abdominal pain was not a feature of his condition, and his symptoms, more often than not, lasted not 'one or two days or up to a week', but for weeks or even months at a time.

2. Lactose Intolerance Syndrome

The metabolism of lactose – a constituent of milk and dairy products – by the body into glucose and galactose, is dependent on the secretion of the enzyme lactase, from the mucosa of the small intestine. Many adults suffer from lactose intolerance which is the result of the lactase level declining with age.

> What happens to the lactose in the intestine of a lactase-deficient person? The lactose is a good energy source for microorganisms in the colon, and they ferment it to lactic acid while also generating methane (CH_4) and hydrogen gas (H_2). The gas produced creates an uncomfortable feeling of gut distension and the annoying problem of flatulence.
>
> The lactic acid produced by the microorganisms is osmotically active and draws water into the intestine, as does any undigested lactose, resulting in diarrhea [a loose, watery stool]. If severe enough, the gas and diarrhea hinder the absorption of other nutrients such as fats and proteins.[62]

This theory can also be discounted, again on the grounds that neither abdominal pain nor diarrhoea were features of Darwin's condition.

3. Infection with *Helicobacter pylori*

Barry J. Marshall and J. Robin Warren of the University of Western Australia were awarded the Nobel Prize for Medicine for their discovery of the bacterium *Helicobacter pylori* and its role in the causation of gastritis and peptic ulcer disease in October 2005. Furthermore, in February 2009, Marshall proposed that *Helicobacter pylori* was the cause of Darwin's chronic ill health.[63]

According to US Professor of Medicine Gerald L. Mandell and others,

> *H. pylori* has been isolated from persons in all parts of the world. The high prevalence and incidence of colonization [i.e. infection of the alimentary tract] among persons in settings where sanitary conditions are suboptimal, including … developing countries, suggests that fecal-oral transmission occurs.[64]

In other words, the infection is probably spread as the result of poor personal hygiene.

> *H. pylori* may cause an acute upper gastrointestinal illness with nausea and upper abdominal pain. Vomiting, burping, and fever may also be present. Symptoms last from 3 to 14 days, with most illnesses persisting less than 1 week.[65]

It was the case that, in both duodenal and gastric ulceration (other than those caused by irritant drugs such as aspirin), *H. pylori* was almost invariably present in the organs concerned.[66]

David A. Warrell, Emeritus Professor of Tropical Medicine at Oxford University, described in detail, how

> *H. pylori* … increases [the] release of the acid-stimulating hormone, gastrin [produced in the duodenum and pylorus, which stimulates the production of gastric acid by the stomach]. The increase in acid secretion directly damages the duodenal mucosa [lining]. Acid hypersecretion also produces gastric metaplasia [abnormal change in the nature of a tissue[67]] in the proximal duodenum. This allows *H. pylori* to colonize the duodenum and produce duodenitis … . These changes result in duodenal ulcers.[68]
>
> Duodenal ulcers typically present with pain that is dull and located in the epigastrium [upper abdomen] or to the right of it over the duodenum itself. It is characteristically relieved by eating, then gets worse when the stomach empties. The pain usually wakes the patient from sleep in the middle of the night and is relieved by eating food, drinking milk, or taking an alkali preparation (antacid). The pain is … episodic with exacerbations lasting a few weeks separated by pain-free periods. These last for several weeks or months and probably reflect spontaneous healing of the ulcer. Patients with duodenal ulcer often have other symptoms such as retrosternal burning and acid regurgitation. Nausea and vomiting are unusual and appetite is preserved.[69]

Colonization of the stomach itself with *H. pylori* may damage its mucosa in a similar way.

> Patients with chronic gastric ulcers tend to be over 40 years old and from lower socioeconomic groups. Epigastric pain is the most frequent symptom. It occasionally radiates to the back if the ulcer is located posteriorly. Food or antacids usually relieve it. It typically occurs in exacerbations lasting for several weeks with symptom-free periods in between. Night pain occurs in a minority of patients with gastric ulcer compared with most of those with

duodenal ulcers. Gastric ulcers quite often produce nausea, anorexia, or weight loss … . Some patients vomit, but many ulcers [i.e. ulcer sufferers] have no symptoms until the patient presents with haemorrhage … .[70]

Once again, in Darwin's case, the absence of abdominal pain, and the frequency and severity of his vomiting, makes it highly unlikely that he suffered from ulcers, and therefore from *Helicobacter pylori* infection.

4. Crohn's disease

This chronic inflammatory disease can affect any part of the alimentary tract, but most commonly the terminal ileum (small intestine). Its onset is usually seen in adolescence, or in those over the age of sixty. Common symptoms are spasms of pain in the abdomen, chronic sickness, loss of appetite, weight loss, and rectal bleeding.

Yet again, the presence of abdominal pain in Crohn's disease rules this out, as far as Darwin is concerned.

Chapter 29

Dr Ralph Colp: Professor Saul Adler: Chagas' Disease

In his exhaustive study entitled 'Darwin's Illness', published in 2008, US psychiatrist Ralph Colp Jr, assesses the validity of the numerous tentative diagnoses which have been made, over the years, in respect of Darwin's condition. Apart from the ones already mentioned, they include toxicity from the taking of snuff and/or from the imbibing of alcohol; gout; chronic neurasthaenia; refractive anomaly of the eyes; pyorrhea of the gums; mental overwork; mild (hereditary) depression; serious anxiety neurosis; chronic brucellosis; malaria; chronic appendicitis; cholecystitis; diaphragmatic hernia; hypoglycaemia; arsenic poisoning; mercury poisoning from the ingestion of medicines containing mercuric chloride; severe allergy, possibly to pigeons, which Darwin studied and bred; hyperventilation syndrome; panic disorder; agoraphobia; dysfunction of the immune system; hypoadrenalism; and systemic lupus erythematosis. However, the only candidates to which Colp gives any credence in respect of Darwin, are psychoneurosis and Chagas' disease.[1]

'Psychoneurosis' is defined as a relatively mild mental illness that is not caused by organic disease, involving symptoms of stress (depression, anxiety, obsessive behaviour, hypochondria) but not a radical loss of touch with reality.[2] To anyone familiar with Darwin, it is obvious that he was a driven man whose overwhelming desire was to explore, discover, and theorize about the natural world. And he became positively morose and depressed at anything that stood in his way in this respect. Work, to him, was not only a tonic, but also a therapy. Said he, in November 1863:

> My chief enjoyment and sole employment throughout life has been scientific work, and the excitement [derived] from such work makes me for the time forget, or drives quite away my daily discomfort.[3]
>
> I have been so steadily going down hill, I cannot help doubting whether I can ever crawl a little up hill again. Unless I can, enough to work a little, I hope my life may be very short; for to lie on [the] sofa all day & do nothing, but give trouble to the best & kindest of wives & good dear children is dreadful.[4]

To journalist, translator, and poet Charles Boner, in early 1870, Darwin wrote, 'You describe the grand scenery of the [Austrian] Tyrol most graphically, and it makes me long to be strong and young again to ramble over the mountains.' This was in reference to Boner's book *Chamois Hunting in the Mountains of Bavaria and in the Tyrol,* published in 1860.[5] These words are hardly those of a psychoneurotic.

Darwin, in his younger days at Cambridge, had not been averse to enjoying himself. However, because of his illness, he said,

> I have therefore been compelled for many years to give up all dinner-parties; and this has been somewhat of a deprivation to me, as parties always put me into high spirits.[6]

Given that psychoneurosis is defined as 'a relatively mild mental illness', then surely Darwin, with his tremendous willpower, would have been able to overcome it. Clearly, there was something far more seriously amiss with Darwin's health.

The time has come, therefore, to consider Dr Colp's other possible diagnosis in respect of Darwin's illness, namely Chagas' disease.

Professor Saul Adler

It was Professor Saul Adler of the Department of Parasitology, Hebrew University, Jerusalem, who first pointed out, in October 1959, the likely cause of the ill health which Darwin experienced all his adult life when he wrote that:

> Darwin's illness, which practically amounted to forty years of invalidism, has given rise to considerable speculation. The doctors who treated him could find no physical explanation for his distressing symptoms and apparently concluded that he was a hypochondriac.[7]

'Hypochondria' is an abnormal anxiety about one's health, especially with an unwarranted fear that one has a serious disease,[8] and even Darwin himself, in a letter to Hooker dated 31 March 1845, stated that 'many of my friends I believe think me a hypochondriac'.[9] Adler, however, introduces a note of caution when he states that 'A purely psychological aetiology for Darwin's illness cannot be accepted as conclusive until all other factors have been eliminated'.[10]

Referring to the voyage of HMS *Beagle*, Adler opined that 'there is nothing of any medical significance prior to the voyage which could throw light on his subsequent illness'. And he quotes Lady Nora Barlow, daughter of Darwin's son Horace, who contrasted Darwin's 'forty years of invalid existence' with his 'youthful vigour … strength and endurance [which were] were well above the average …'.[11] Adler himself noted that

> Darwin was a dedicated geologist and throughout his whole life maintained the keenest interest in this subject. Nevertheless, at the age of thirty-three, he was compelled to abandon field work in a favourite subject because he found by experience that the physical effort it entailed exhausted him.[12]

Adler now proceeds to quote from Darwin's *Voyage of the Beagle*, and in particular the March 1835 episode at Luxan in the Argentinian province of Mendoza, where Darwin, by his own admission, was attacked by the 'great black bug of the Pampas'. In Darwin's own words:

> It is most disgusting to feel soft wingless insects, about an inch long, crawling over one's body. Before sucking they are quite thin, but afterwards they become round and bloated with blood … . One which I caught at Iquique [Chile] (for they are found in Chile and Peru) was empty. When placed on a table, and though surrounded by people, if a finger was presented, the bold insect would immediately protrude its sucker, make a charge, and if allowed, draw blood. No pain was caused by the wound. It was curious to watch its body during the act of sucking, as in less than ten minutes it changed from being as flat as a wafer to a globular form. This one feast, for which the *Benchuca* [triatome bug – see below] was indebted to one of the officers, kept it fat during four whole months; but after the first fortnight, it was quite ready to have another suck.[13]

Adler identifies the 'bug' as *Triatoma infestans* which, in turn, is the 'most important vector [carrier] of *Trypanosoma cruzi* [a single-celled parasitic protozoan with a trailing flagellum – tail],[14] the causative agent of Chagas' disease in the Argentine, Chile and parts of Brazil'.[15] He continued,

> The province of Mendoza has a relatively high incidence of Chagas' disease, and according to South American colleagues with whom I discussed this problem at the recent congress on Chagas' disease held in Rio de Janeiro during July 5–12 [1959] as many as 60 per cent of the population in parts of Mendoza give a positive complement-fixation test for T. [*Trypanosoma*] cruzi … .

The complement fixation test was used, at that time, to diagnose infectious diseases by detecting the presence of a specific antibody or antigen in the patient's blood serum.

> [Also] … as many as 70 per cent of specimens of *Triatoma infestans* are infected with the trypanosome. Darwin was therefore definitely exposed to infection on at least one occasion. It is highly probable that he was also exposed on other occasions … .

> We must also bear in mind that Chagas' disease has a very wide distribution in South America from Chile to Mexico (recently a few cases have been recorded in Texas) and the province of Mendoza is an area of relatively high infestation.
>
> The incident in Luxan cannot, however, explain Darwin's previous seven weeks severe illness during September and October 1834 which confined him to bed in Valparaiso and commenced during the last week of a six weeks journey. Unfortunately, no clinical details of this episode are available.

However, said Adler, according to Scottish anatomist and anthropologist Sir Arthur Keith, typhoid fever was 'a very probable diagnosis' for this unexplained illness.[16] Adler continued:

> The complications and sequelae of Chagas's disease have been studied in detail by some of the ablest South American pathologists and clinicians, particularly in Brazil, the Argentine, Chile and Uruguay, and considerable literature on this subject is now available. Particular attention has been paid to the clinical and pathological aspects of the myocarditis [inflammation of the heart muscle] which appears in some victims of the disease. Darwin's exhaustion after physical effort can well be explained on the basis of an infection with *T. cruzi*.

Adler concludes that

> It is obviously impossible to *prove* that Darwin was a victim of Chagas' disease but two points cannot be overlooked: (1) his symptoms can be fitted into the framework of Chagas' disease at least as well as into any psychogenic [having a psychological origin or cause rather than a physical one][17] theory for their origin; (2) it is possible to pin-point with certainty a definite incident on March 25, 1835, during which he was exposed to optimal conditions for infection with *T. cruzi*.[18]

Chagas' disease

Chagas' disease, otherwise known as New World or South American trypanosomiasis, is a disease principally of northern, north-western and central-south America, and it was Carlos Chagas (1879–1934) who discovered the disease that now bears his name. The son of a Brazilian coffee grower, Chagas graduated from the school of medicine at Rio de Janeiro in 1902 and was awarded his doctorate in medicine the following year whilst working at the new Medical Research Institute founded by physician, bacteriologist, epidemiologist and public health officer Oswaldo Cruz. Having left the institute for the port of Santos, São Paulo (200 miles to the south-west), to help combat a malaria epidemic, he returned in 1906, and the following year was sent to the city of Lassance (450 miles to the north) to combat another malaria outbreak. It was during his two-year sojourn at Lassance that he studied the behaviour of blood-sucking insects, known as *triatomines*. These 'bugs' are commonly found in the cracks and crevices of the walls and ceilings of primitive rural dwellings in which they hide by day, and emerge at night to suck the blood of those sleeping in the house, whether human beings or domestic animals.

When Chagas examined the intestinal contents of the *triatomes* he discovered that they harboured a new species of the *Trypanosoma* genre.[19] Suspecting that the *triatome* 'bugs' might be instrumental in transmitting their trypanosome parasites to man and other vertebrates, Chagas

> therefore sent [a collection of] bugs to our Institute where the director, Dr Oswaldo Cruz, tried to infect a monkey of the species *Callithrix penicillata*, by having it bitten by several examples of the hemipteran [*triatomes*]. Twenty or thirty days after the bite, large numbers of trypanosomes were found in the peripheral blood of the monkey, with morphology that was entirely distinct from any known species of the genus *Trypanosoma*.[20]

Chagas named this new parasite *Trypanosoma cruzi*, in honour of Oswaldo Cruz.

In 1912 E. J. Alexander Brumpt, Professor at the Faculty of Medicine and Director of the Parasitology Laboratory in Paris, challenged Chagas' view that the disease was transmitted by the *bite* of the triatomine. He suggested, instead, that in most cases it was the faeces of infected bugs which transmitted the disease via the mucous membranes of humans (or animals). In other words, a bug lands on a person, deposits its faeces on the skin, the person then touches or scratches the spot, his hands become contaminated, he then touches his mouth, which gives the trypanosome easy access to the body via the mouth's lining. (That this was the usual mode of transmission was later confirmed by Chagas' disciple at the Oswaldo Cruz Institute, Dr Emmanuel Dias.)[21] It has since been discovered that the disease can also be spread by consuming the uncooked meat of animals infected with the trypanosome, or by consuming food contaminated by the faeces of infected bugs.

The course of the disease

Between the time of a person contracting Chagas' disease and the onset of its clinical symptoms, a delay of between four and forty years is possible.

> The early or acute phase of Chagas' disease is frequently without symptoms, but is occasionally fatal: there may be an initial lesion at the site where *T. cruzi*

has entered the body. The acute phase may resolve in a symptomless, indeterminate period, which can be life-long, but can progress into chronic Chagas' disease with heart abnormalities or intestinal malfunction and enlargement, especially of the oesophagus and colon, with damage to the nerve supply of these organs.[22]

The pathology of the disease

In 1968 Austrian-born Fritz Köberle, Chairman (from 1953) of the Department of Pathology at the Medical School of Ribeirao Preto, Brazil, described how, in patients with Chagas' disease, there was a significant loss of ganglion (assemblage of nerve) cells from the parasympathetic autonomic nervous system (the so-called Auerbach's plexus of nerve fibres and ganglia, which acts upon the muscle of the gastrointestinal tract and controls its motility). This explains why the oesophagus and/or colon of patients infected with *Trypanosoma cruzi* may lose their normal muscle tone, fail to perform peristalsis, and become dilated. (This is of particular significance as far as Darwin's illness is concerned, as will be seen.) The colon may be affected in a similar way and also the heart, leading to chronic heart disease, heart failure, cardiomegaly (enlargement of the heart), arrhythmias (irregularities of the pulse) and thromboembolism; all of which are potentially life-threatening conditions.[23]

Arguments against the Chagas' disease hypothesis

i. Wikipedia

The online website Wikipedia, under the heading 'Evidences against the Chagas hypothesis', lists the following objections to Professor Saul Adler's theory in respect of Darwin.[24]

(The author of this volume's comments appear, in italics, after each entry):

'Darwin died at a relatively old age for his time (seventy-three years old)'.

The modern consensus of opinion is that the presence of chronic Chagas' disease (in the absence of other factors) reduces the patient's lifespan by up to one decade only. Darwin's father Robert, lived to be eighty-two.

'The symptoms abated as he aged, which is not typical for the disease, where age exacerbates the symptoms.'

Untrue. His symptoms did not *abate as he aged.*

'He did not seem to have several of the pathological damages present at [i.e. in patients with] chronic Chagas' disease, such as megacolon and megaesophagus.'

As the technique of clinical biopsy was not available in Darwin's time, and in the absence of post-mortem evidence, this statement is impossible to verify. What is true is that Darwin's digestive symptoms were typical *of megaoesophagus.*

'Some of the symptoms, such as tachycardia, fatigue and tremors, were already present before the *Beagle* voyage.'

Possibly, but not to the extent that they impeded Darwin in any significant way. Had they done so, he would have been unable to undertake the rigours of the Beagle *voyage.*

'The numerous partial exacerbations and remissions [experienced by Darwin] are unusual in Chagas' disease.'

This is not true, at any rate as far as Darwin's digestive symptoms are concerned, for according to Professor Antonio R. L. Teixeira (Director of the Chagas' Disease Multidisciplinary Research Laboratory, Faculty of Medicine, University of Brasilia, Brazil) and others, the digestive symptoms of the disease 'evolve during periods of dysphagia [difficulty or discomfort in swallowing[25] and are], followed by long periods during which symptoms are absent', and so on in a continual cycle.[26]

'The incidence of trypanosome-infected benchucas [triatome bugs] in Mendoza, Argentina (which had a colder climate) where Darwin reported the bite, is very low.'

It is true, according to data provided by the World Health Organization, that the Andes Mountains are a trypanosome-free zone[27] *(presumably because here, there are few animals or beasts for the triatomes to feed off), but the same is not true of the foothills of the Andes, where Luxan is situated and where Darwin had personally encountered the bugs.*

'No other members of *Beagle*'s crew who accompanied Darwin in his land trip showed signs of a similar disease.'

This cannot be said with certainty. Darwin lost touch with the majority of the Beagle*'s crew members; so if they had become ill, he would not necessarily have known about it. Also, it is unlikely that any physician of the time would have recognized the significance of the symptoms as it would be another seven decades before the nature of the disease was elucidated. However, it is possible that one or more of the small* *proportion of the members of the* Beagle*'s crew who ventured into the interior,* did *become infected with trypanosomes. (Recent estimations of Chagas' disease prevalence in the rural populations of Argentina, for example, show values ranging from 25 per cent to 45 per cent.)*[28]

However, according to Tiexeira et al, only 'one third of all individuals with indeterminate infections ['indeterminate' meaning the period following infection that precedes the appearance of clinical signs and symptoms] will develop chronic Chagas' disease]'. [29]

ii. Professor John A. Hayman
In an article entitled 'Darwin's Illness Revisited', published in December 2009,[30] Professor John A. Hayman, stated:

> Chagas' disease is rejected for several reasons: exposure [i.e. to the infecting agent] was too brief and Darwin had symptoms before he sailed [on the *Beagle*]. His tolerance of exercise was good and despite being examined by several eminent doctors he showed no evidence of organic disease.

These propositions of Professor Hayman's may be taken one by one and largely refuted, for the following reasons:

1. In March/April 1835, Darwin spent a month travelling through regions of Chile and Argentina where what later became known as Chagas' disease was endemic. He also made other protracted excursions into Chagas-infected regions of Central South America.

2. Before HMS *Beagle* sailed, Darwin complained of 'palpitation and pain about the heart'. These symptoms were of a transient nature and they in no way resembled the chronic symptoms which would plague him all his life.

3. Although robust in health for most of the *Beagle* voyage, Darwin subsequently suffered for the remainder of his life from extreme lassitude.

4. It is true that Darwin was examined by a host of eminent doctors but, unfortunately, none of them left behind for posterity any case notes relating to their patient. Nonetheless, the limitations of medical science at the time would have precluded them from being able to make a definitive diagnosis of either megacolon or megaoesophagus in a patient.

Chapter 30

Darwin, Emma, and God

Emma regrets that Darwin does not share her religious views

The following letter, written by Emma to Darwin in late November 1838, two months prior to their marriage, indicates that, as far as religion was concerned, there were significant differences between them.

> When I am with you I think all melancholy thoughts keep out of my head but since you are gone [i.e. returned to Shrewsbury] some sad ones have forced themselves in, of fear that our opinions on the most important subject should differ widely. My reason tells me that honest & conscientious doubts cannot be a sin, but I feel it would be a painful void between us. I thank you from my heart for your openness with me & I should dread the feeling that you were concealing your opinions from the fear of giving me pain. It is perhaps foolish of me to say this much but my own dear Charley, we now do belong to each other & I cannot help being open with you. Will you do me a favour? yes I am sure you will, it is to read our Saviour's farewell discourse to his disciples which begins at the end of the 13th Chap[ter] of [the Gospel according to St] John. It is so full of love to them & devotion & every beautiful feeling. It is the part of the New Testament I love best.[1]

This was a reference to Christ's words, 'A new commandment I give unto you. That ye love one another; as I have loved you.'[2]

The following January, Emma returned to the subject.

> There is only one subject in the world that ever gives me a moment's uneasiness & I believe I think about that very little when I am with you & I do hope that though our opinions may not agree upon all points of religion we may sympathize a good deal in our *feelings* on the subject.[3]

In his autobiography Darwin indicates that his feelings about Christianity had become polarized during the two years or so prior to his marriage to Emma. Said he,

> disbelief crept over me at very slow rate, but was at last complete. The rate was so slow that I felt no distress, and have never since doubted even for a single second that my conclusion was correct. I can indeed hardly see how anyone ought to wish Christianity to be true; for if so the plain language of the text seems to show that the men who do not believe, and this would include my Father, Brother and almost all of my friends, will be everlasting punished. And this is a damnable doctrine.[4]

In February Emma told Darwin:

> Your mind & time are full of the most interesting subjects & thoughts of the most absorbing kind, viz following up yr own discoveries – but which makes it very difficult for you to avoid casting out as interruptions other sorts of thoughts which have no relation to what you are pursuing, or to be able to give your whole attention to both sides of the question.
>
> May not the habit in scientific pursuits of believing nothing till it is proved, influence your mind too much in other things which cannot be proved in the same way, & which if true are likely to be above our comprehension.[5]

Reading between the lines, Emma appears to be regretting the fact that her husband Charles, has no time for the Christian doctrine on the grounds that it is scientifically unverifiable. As for Darwin, he agonized over Emma's letter in the knowledge that his scientific research was leading him to conclusions which distressed his wife greatly, and at a later date, he added to it the words 'When I am dead, know that many times, I have kissed & cryed over this.'[6]

When Fox asked Darwin to become godfather to his forthcoming child, he received the following reply.

> I conceive myself bound to tell you, that we have not had Godfathers or Godmothers to our children, not from any objection to their having such – but as we should in that case have been obliged to have stood proxies & we both disliked the statement of believing anything for [on behalf of] another.[7]

In other words, it was Darwin's view, and according to him the view of his wife Emma also, that it was for the individual, whether adult or child, to decide what, if any, religion he or she was to adopt. Nevertheless, he told Fox, 'our children are baptized'.[8]

Emma is described as a 'sincerely religious' person who went regularly to church and took the Sacrament (i.e. participated in the ceremony of Holy Communion). She read the *Holy Bible* with her children and taught them 'a simple Unitarian Creed', though they were 'baptized and confirmed in the Church of England. In her youth, religion must have largely filled her life ....'[9]

Emma told her aunt Jessie Sismondi, on 27 August 1845, that Darwin had just completed his journal (*Journal of Remarks*, commonly known as *The Voyage of the Beagle*). She herself had just finished reading an autobiography entitled *The Life of St Blanco the Martyr.* (The correct title of this volume is *The Life of the Rev. Joseph Blanco White*, published in 1845.)

Joseph Blanco White

Spanish theologian and poet Joseph Blanco White, christened José María Blanco y Crespo, was born in Seville, Spain in 1775 to an English father and a Spanish mother. Having studied for the Catholic priesthood at the University of Seville he was ordained priest in 1800. In that same year, at the age of twenty-five, he was elected Rector of Seville's Dominican Colegio Mayor de Santo Tomás de Aquino.[10] However, he became disillusioned and developed a

> growing conviction that Christianity itself, in its Catholic, Spanish and specifically Sevillan form, was a baleful force which, manipulated by a privileged theocracy, blighted human happiness and obstructed social progress.[11]

In 1805 White arrived in Madrid. However, because of his religious doubts, he decided to flee Spain and, in February 1810, he sailed from Cadiz to Falmouth on the English frigate *Lord Howard*, never to return to his native land.[12]

Having read and been inspired by the Reverend William Paley's *Natural Theology*, White converted to Anglicanism.[13] In 1826 Oriel College, Oxford bestowed upon him the degree of Honorary Master of Arts and admitted him as a fellow of the college. He became a friend of theologian Richard Whateley (who was also a fellow of Oriel) and when, in 1831, the latter became Archbishop of Dublin, White became tutor to his family. However, he subsequently 'jumped ship' once again, by embracing Unitarianism and relocating to Liverpool, which had a thriving Unitarian community. White died on 20 May 1841 and was buried at Liverpool's Renshaw Street (Unitarian) Chapel.[14]

Emma would have been particularly gratified that White, having sampled both Catholicism and Anglicanism, finally chose to become, like her and her family, a Unitarian. Of *The Life of St Blanco the Martyr*, she

> would advise every scientific man who is preparing a new edition in any rapidly progressive branch of science, in which he has launched many new speculations and theories, to read … and to be grateful that in the department which he has to teach he is not pledged to retain forever the same views, or that the slightest departure from them need not entail on him the penalty of the loss of nearly all worldly advantages, domestic ties, and friendships. How ashamed

> ought every lover of truth to feel if mere self-love or pride makes him adhere obstinately to his views, after seeing the sacrifices which such a man [as White] was ready to make for what he believed to be the truth. This is the moral I draw from the book.[15]

Here then was Emma, in a clear allusion to her husband Charles, echoing Paley, by implying that 'truth' has a greater validity when uttered by those who have suffered or made sacrifices for it. What an irony this was, in view of the furore which had greeted the publication of Darwin's own magnum opus *Origin*, and the suffering which he himself had been forced to endure, having told the truth, as he saw it, in relation to his scientific discoveries!

How was life begun, if not by the Creator?

Darwin was aware, from his early studies of geology, that the Earth was far older than 6,000 years – a figure deduced from genealogical studies, based on Adam and Eve and their descendants, these two persons being, according to the *Holy Bible*'s Book of Genesis, the first people to inhabit the Earth. (Earth is, in fact, estimated to be about 4.50 billion years old). And, as soon as he saw his first fossil, he would have realized that the Biblical account of creation, as detailed in the *Old Testament*'s Book of Genesis, was simply a figment of its author's imagination. Nevertheless, he was prepared to believe that a Creator had 'breathed life' into the earliest living creature, or creatures.[16] Later, however, he appears to suggest that life may have originated without any necessity for the presence of God. This is indicated in a letter which he wrote to Joseph Hooker on 1 February 1871.

> It is often said that all the conditions for the first production of a living organism are now present, which could ever have been present. But if (& oh what a big if) we could conceive in some warm little pond with all sorts of ammonia & phosphoric salts, light, heat, electricity, &c., present, that a protein compound was chemically formed, ready to undergo still more complex changes, [then] at the present day such matter would be instantly devoured or absorbed, which would not have been the case before living creatures were formed.[17]

In other words, if living matter had, in Darwin's day, emerged from some 'warm little pond', then the likelihood is that man or animals would have destroyed it before it had a chance to develop. In the above letter to Hooker, Darwin had, wittingly or unwittingly, anticipated that further research would be done into this subject during the following century, by which time the science of chemistry would have become farther advanced.

The creation of life artificially

Abiogenesis is the theory that, if conditions are appropriate, life can arise spontaneously from non-living molecules. Darwin had once opined to Wallace, as already mentioned, that

> if it could be shown that life had generated itself spontaneously then this 'would be a discovery of transcendent importance'.[18]

In 1953, Harold C. Urey of the University of Chicago and his twenty-three-year-old graduate student Stanley L. Miller, conducted an experiment in which they simulated conditions believed to be present at the time life on Earth began. In the experiment, electric sparks (i.e. simulating lightning) were continually passed through a flask containing water (heated), methane, ammonia, and hydrogen (but *not oxygen*, as this did not exist in the atmosphere before plant life began). Two weeks later, Miller and Urey observed that two per cent of the carbon present was now in the form of amino acids – organic compounds which occur naturally in plant and animal tissues and are the basic constituents of proteins.[19]

When the US Professor Carl Sagan (1934–96), astronomer and astrophysicist, subsequently conducted a similar experiment,

> a rich collection of complex organic molecules, including the building blocks of the proteins and the nucleic acids [were created]. Under the right conditions, these building blocks assemble themselves into molecules resembling little proteins and little nucleic acids. These nucleic acids can even make identical copies of themselves.[20]

This research gives an indication as to how life on Earth began.

Finally, in a letter to US clergyman and writer Francis E. Abbot, Darwin declared, 'I can never make up my mind how far an inward conviction that there must be some creator or first cause is really trustworthy evidence'.[21] In other words, simply to have faith in such a notion was, for Darwin, not enough.

What did Darwin really believe?

That the notion of a Divine Designer was problematical

To Asa Gray on 22 May 1860, Darwin continued to agonize over the apparent contradiction between his theory of evolution and the Christian doctrine of Creator/Designer.

> With respect to the theological view of the question of living things; this is always painful to me. I am bewildered. I had no intention to write atheistically. But I own that I cannot see, as plainly as others do, & as I sh^{d} wish to do,

> evidence of design & beneficence on all sides of us. There seems to me too much misery in the world. I cannot persuade myself that a beneficent & omnipotent God would have designedly created the Ichneumonidae [a family within the insect order hymenoptera] with the express intention of their [larvae] feeding within the living bodies of caterpillars, or that a cat should play with mice. Not believing this, I see no necessity in the belief that the eye was expressly designed. On the other hand I cannot anyhow be contented to view this wonderful universe & especially the nature of man, & to conclude that everything is the result of brute force. I am inclined to look at everything as resulting from designed laws, with the details, whether good or bad, left to the working out of what we may call chance. Not that this notion *at all* satisfies me. I feel most deeply that the whole subject is too profound for the human intellect. A dog might as well speculate on the mind of [Sir Isaac] Newton. Let each man hope & believe what he can.
>
> Certainly I agree with you that my views are not at all necessarily atheistical. The lightning kills a man, whether a good one or bad one, owing to the excessively complex action of natural laws, a child (who may turn out an idiot) is born by [the] action of even more complex laws, and I can see no reason, why a man, or other animal, may not have been aboriginally produced by other laws; & that all these laws may have been expressly designed by an omniscient Creator, who foresaw every future event & consequence. But the more I think the more bewildered I become … .[22]

A decade later, on 12 July 1870, Darwin told Hooker:

> Your conclusion that all speculation about preordination [where an outcome is decided or determined beforehand[23]] is [an] idle waste of time is the only wise one: but how difficult it is not to speculate. My theology is a simple muddle: I cannot look at the universe as [being] a result of blind chance, yet I can see no evidence of beneficent design, or indeed of design of any kind in the details.

In other words, the problem remained unresolved. As for 'spontaneous generations [this] seems almost as great a puzzle as preordination ….'[24]

That the universe is subject to natural laws with which the Creator does not interfere

Darwin wrote to Lyell on 17 June 1860 to say:

> One word more upon the 'Deification' [deify – 'to make godlike in character'] of Natural Selection … . No astronomer in showing how movements of Planets are due to gravity, thinks it necessary to say that the law of gravity was

> designed [i.e. by God, in order] that planets sh^{d} pursue the courses which they pursue. I cannot believe that there is a bit more interference by the Creator in the construction of each species, than in the course of the planets.
>
> It is only owing to Paley & Co, as I believe, that this more special interference is thought necessary with living bodies. But we shall never agree, so do not trouble yourself to answer.[25]

Paley had argued that 'the organization and adaptations of living organisms indicated the existence of an intelligent creator'.[26]

Here, Darwin appears to be saying that the Creator does not involve himself with determining the structure of every 'organic being'; this structure having evolved according to 'general laws' – and he includes 'natural selection' as one of these laws.

Science and religion: shall ere the twain meet?

Darwin received a letter, on 14 December 1866, from writer and educator Mary E. Boole, in which the following searching questions were asked:

> Do you consider the holding of your Theory of Natural Selection, in its fullest & most unreserved sense, to be inconsistent … with the following belief, viz:
>
> That knowledge is given to man by the direct Inspiration of the Spirit of God.
>
> That God is a personal and Infinitely good Being.
>
> That the effect of the action of the Spirit of God on the brain of man is *especially* a moral effect.
>
> And that each individual man has, within certain limits, a power of choice as to how far he will yield to his hereditary animal impulses, and how far he will rather follow the guidance of the Spirit Who is educating him into a power of resisting those impulses in obedience to moral motives.
>
> The reason why I ask you is this. My own impression has always been, – not only that your theory was quite *compatible* with the faith to which I have just tried to give expression, – but that your books afforded me a clue which would guide me in applying that faith to the solution of certain complicated psychological problems which it was of practical importance to me, as a mother, to solve [Boole was the mother of five daughters]. I felt that you had supplied one of the missing links, – not to say the missing link, – between the facts of Science & the promises of religion. Every year's experience tends to deepen in me that impression.
>
> But I have lately read remarks, on the probable bearing of your theory on religious & moral questions, which have perplexed & pained me sorely.
>
> At the same time I feel that you have a perfect right to refuse to answer such questions as I have asked you. Science must take her path & Theology hers,

> and they will meet when & where & how God pleases, & you are in no sense responsible, for it, if the meeting-point should be still very far off.[27]

In her preface to Mary Boole's *Collected Works*,[28] US educationist Ethel S. Dummer, gives a clue as to what may have prompted the author to write to Darwin in this way, when she describes how Boole had delivered a series of lectures to a group of London mothers 'who, finding their religion threatened by Darwin's new theories, sought Mrs Boole's philosophic wisdom' on the subject.[29]

Darwin's response to Mary Boole, given on 14 December 1866, reads:

> Dear Madam.
> It would have gratified me much if I could have sent satisfactory answers to y[r]. questions, or indeed answers of any kind.... I may however remark that it has always appeared to me more satisfactory to look at the immense amount of pain & suffering in this world, as the inevitable result of the natural sequence of events, i.e. general laws, rather than from a direct intervention of God though I am aware this is not logical with reference to an omniscient Deity. Your last question seems to resolve itself into the problem of Free Will & Necessity which has been found by most persons insoluble.
>
> P.S. I am grieved that my views should incidentally have caused trouble to your mind but I thank you for your Judgement & honour you for it, that theology & science should each run its own course & that in the present case I am not responsible if their meeting point should still be far off.[30]

The above correspondence reaches to the very heart of the question: is religion (in this case Christianity) compatible with Darwinism? But, instead of being drawn into furnishing a definitive answer, Darwin, who was as exemplary a scientific investigator as ever lived, evidently prefers to keep an open mind on this highly contentious subject. On the one hand, he appears to give an implicit acknowledgment that, as yet, there is no convergence of opinion (i.e. compatibility) between science and religion, whilst on the other, he declares that such a convergence cannot be ruled out at some future date – presumably when science has shed more light, not only on the mechanism of evolution, but also on that of creation itself. Or did he, privately and in his heart of hearts, secretly believe that such a reconciliation could never come about?

To conclude, the root cause of Darwin's dilemma was that whereas his Theory of Evolution was based on a vast amount of scientific evidence, painstakingly accumulated, tested, and evaluated over many years, the doctrine of Christianity, as far as he could determine, was unable to draw on one single scientific fact to back it up.

Which holds sway: the laws of a benificent God, or the laws of nature?

In 1867, when Darwin completed his great work *The Variation of Animals and Plants under Domestication* (*Variation*), he declared to Hooker:

> I finish my Book … by a single paragraph answering, or throwing doubt, in so far as so little space permits on Asa Gray's doctrine that each variation has been specially ordered or led along a beneficial line. It is foolish to touch such subjects, but there have been so many allusions to what I think about the part which God has played in the formation of organic beings, that I thought it shabby to evade the question.[31]

The paragraph referred to above reads:

> However much we may wish it, we can hardly follow Professor Asa Gray in his belief 'that variation has been led along certain beneficial lines,' like a stream 'along definite and useful lines of irrigation'. If we assume that each particular variation was from the beginning of all time preordained, the plasticity [the quality of being easily shaped or moulded][32] of organization, which leads to many injurious deviations of structure, as well as that redundant [no longer necessary or useful; superfluous][33] power of reproduction which inevitably leads to a struggle for existence, and, as a consequence, to the natural selection or survival of the fittest, must appear to us superfluous laws of nature. On the other hand, an omnipotent and omniscient Creator ordains everything and foresees everything. Thus we are brought face to face with a difficulty as insoluble as is that of free will and predestination [predestine – to ordain in advance by divine will].[34]

In *Variation* Darwin elaborated further upon this theme. Was it conceivable, he enquired sceptically, that the

> Creator … specially ordained for the sake of the breeder each of the innumerable variations in our domestic animals and plants; many of these variations being of no service to man and not beneficial, [but instead were] far more often injurious to the creatures themselves?[35]

Here, Darwin is objecting to Asa Gray's view that variation was preordained by God along certain beneficial lines, on the grounds that a), it does not explain the existence of non-beneficial variations and b), it is a negation of his [Darwin's] theory of evolution by natural selection.

The Reverend William Paley

In all the circumstances it is surprising that Darwin retained his fondness for the works of the Reverend William Paley, which he had studied whilst an undergraduate at Cambridge. This fondness, he revealed when, at the age of fifty, he wrote to banker, politician and naturalist Sir John Lubbock, a neighbour of his, of High Elms, Down, to say, 'I hardly think I ever admired a book more than Paley's natural theology [*Natural Theology, or Evidence of the Existence and Attributes of the Deity*, published in 1802]. I could almost formally have said it by heart.'[36] The impression gained is that Darwin would dearly have loved to embrace the teachings of Paley, but reason dictated that he simply could not bring himself to do so.

* * *

Emma told Darwin, in June 1861, in respect of his chronic illness,

> My heart has often been too full to speak or take any notice[.] I am sure you know I love you well enough to believe that I mind your sufferings nearly as much as I should my own & and I find the only relief to my own mind is to take it as from God's hand, & to try to believe that all suffering & illness is meant to help us to exalt our minds & to look forward with hope to a future state.

In other words, Emma's love for her husband had remained undimmed, despite his failure to respond to her religious overtures.

> When I see your patience, deep compassion for others[,] self-command & above all gratitude for the smallest thing done to help you I cannot help longing that these precious feelings should be offered to Heaven for the sake of your daily happiness … . I often think of the words 'Thou shalt keep him in perfect peace whose mind is stayed [i.e. fixed] on thee [a quotation from the Book of Isaiah]'.[37] It is feeling & not reasoning that drives one to prayer.[38]

That year, in the July edition of *Macmillan's Magazine*, there appeared an article written by Frances Julia Wedgwood (daughter of Emma's brother Hensleigh and his wife Frances) entitled 'The Boundaries of Science, A Second Dialogue'. Having read it, Darwin told its author,

> I must confess that I could not clearly follow you in some parts, which probably is in [the] main part due to my not being at all accustomed to metaphysical [i.e. abstract or supernatural] trains of thought.[39]

In March 1869 Darwin wrote to the lawyer Vernon Lushington, in respect of French philosopher and social theorist Auguste Comte (1798–1857). Comte had founded the Société Positiviste in 1848, which was devoted to the 'Cult of Humanity' – his philosophy being described by himself as 'Positivism'. This asserts that the only authentic knowledge is that which can be scientifically verified, or which is capable of logical or mathematical proof. Comte's Positivism therefore demanded a rejection of theism (the belief in a god, or gods). Said Darwin to Lushington:

> No doubt the law of progress from the theological to the positive point of view, is an important one, [and] if true [is a matter] on which I cannot judge, & I sh[d] think the attempt to reduce the social system to a science state seems important.[40]

Here, Darwin admits to the possibility that a law exists which requires a 'theologically-based state' to be replaced by 'science-based state'. But if this is the case, then it begs the question, what place is there for God in such a world?

In January the following year, Emma, having read the proofs of Darwin's most recent book *The Descent of Man and Selection in Relation to Sex*, said frankly, 'I think it would be very interesting, but that I shall dislike it very much as again putting God further off.'[41]

To Huxley in September Darwin wrote, 'God forgive me for writing so long & egotistical a letter … .'[42] Did he mention God purely out of habit, or did he, after all, still retain in his mind some notion of a deity?

Darwin told Francis Abbot in November:

> Now I have never systematically thought much on Religion, in relation to Science, or on morals in relation to Society, & without steadily keeping my mind on such subjects for a *long* period, I am really incapable of writing anything worth sending to the Index. Many years ago I was strongly advised by a friend never to introduce anything about religion – in my works, if I wished to advance science in England; & this led me not to consider the mutual bearings of the two subjects. Had I foreseen how much more liberal the world would become, I should perhaps have acted differently.[43]

Abbot was a US clergyman and philosopher and founder and editor of the *Index*, a weekly paper devoted to promoting 'free religion'. This advocates that all dogma and reliance on scriptures or creeds should be rejected, and that it was up to the individual to seek out the truth for himself.

Chapter 31

Religions: Their Creation and Evolution

It is interesting to speculate as to what conclusions Darwin would have drawn, had he been alive today and chosen to conduct research into religion with the same rigour and objectivity as he had conducted his studies of natural history. Firstly, he would have observed that there are countless 'species' of religion, not only those in vogue today, but many others which are no longer practised and have become 'extinct'.

Of the present-day religions, many appear to have been spontaneously created; whereas others have 'evolved' from those already in existence, sometimes as offshoots and sometimes as a result of violent schisms (a schism being a split or division between strongly opposed sections or parties, caused by differences in opinion or belief[1]).

Hinduism. This religion claims to be the world's oldest, and is based on the concept of a supreme spirit – Brahman – the creative force who brought the universe into being. Depending on their particular sect, Hindus worship one or other of their three major gods: Brahma, Vishnu or Shiva. In addition, there are many thousands of personal gods which any Hindu is free to worship, if he or she so chooses.

Hindus believe that, whereas the human body is destructible, the soul is immutable and takes on different lives in a cycle of birth, death and reincarnation. The sum of a person's actions – or 'Karma' – in one life, determines whether he or she will face the punishments of Hell, or enjoy Heaven as a reward for good deeds. A bad Karma might be rewarded by rebirth as a lower animal; a good Karma by rebirth into a good family with the prospect of a joyous lifetime.

Most Hindus believe that the soul of every person – the 'atman' – is eternal, and that the goal of life is to realize that this is identical to that of the Brahman, and to hope that, when the cycle of life and death is ended, their atman will be absorbed into the Brahman.

Ancient Egyptian religion. Egypt's Early Dynastic Period commenced in about 3,000BC with the unification of the country, though there is evidence of religious activity even prior to this date. In excess of sixty deities existed, including the Sun god Ra; the creator god Amun; and the mother goddess Isis, one or other of which predominated at different periods of history.

Religious life centred on the pharaoh – a human being who was believed to be descended from the gods. He acted as intermediary between the gods and his people, and was obliged to make ritual offerings in order to sustain the gods, so that they might continue to maintain order in the universe.

Egyptians believed in the afterlife, though in early times only the pharaoh and those noblemen whom he particularly favoured could achieve it. Later, however, it was believed that the afterlife was available to all.

Confucianism. Confucius (551–479BC) was a Chinese philosopher who emphasized the importance of justice, sincerity, and morality. However, by the time of the Han dynasty (206BC–AD220), 'Confucianism' had evolved in such a way that both the emperor and the hierarchy of officialdom were regarded as having been divinely appointed; they worshipped the Sun, Heaven, Earth and other 'nature gods', whereas 'ancestor worship' was practised by the people.

Buddhism. Buddha (c.563–483BC) was born the son of a king in northern India. Having achieved 'enlightenment', he proceeded to teach that existence necessarily involves suffering; that the principal cause of suffering is desire, and that the suppression of suffering can be achieved by the suppression of desire. If so, then a state of *Nirvana* is achieved, which is individual extinction and absorption into the supreme spirit.

Christianity. Jesus Christ (c.7–2BC–c.AD30)[2], who was born in Bethlehem in what is now the State of Israel, allegedly performed miraculous healings and exorcisms before being executed by the Romans. Shortly afterwards he was 'resurrected' from the dead and ascended into Heaven. According to St Matthew (apostle and gospel writer), Christ was 'the Son of the living God',[3] who, according to St John (also an apostle and gospel writer), declared, 'I am the way, the truth, and the life: no man cometh to the Father, but by me.'[4] In other words, a belief in Christ was necessary in order to gain access to Heaven where Christ's father, God, dwelt.

Islam. Mohammed (or Muhammad, c.AD570–c.AD632) was an Arab prophet from Mecca – capital of what is now Saudi Arabia. In about the year 610 he claimed that the *Koran* – the sacred book of Islam – had been divinely revealed to him by the angel Jibra'el (Gabriel). Mohammed was the messenger for, and voice of Allah, the creator of all things (including the *Koran*).

* * *

Although the basic tenets of only a handful of the world's religions have been outlined above, from the samples certain patterns are already becoming discernible.

Firstly, God is invisible to the human senses. However, believers often choose to depict god as an object familiar to them in their everyday lives, for example:

Hindus. Brahma is traditionally portrayed with four heads and four arms. The god, Ganesha, has the head of an elephant, four arms, and two legs. The goddess Kali, is depicted in human form and has either four or ten arms.

Egyptians. Gods are typically depicted in humanoid form, occasionally with variations. For example, the god Anubis has the head of a jackal; Sobek takes the form of a crocodile, or of a man with the head of a crocodile; Bastet of a cat.

Buddhists. Buddha is depicted as a human being, as is Kwan Yin, goddess of compassion.

Greeks. Their gods Zeus (the king of the gods) and Aphrodite (goddess of love and beauty), together with hundreds of other deities, all have humanoid forms.

Romans. The same is true of their gods as for the Greeks.

Christians. The Christian god (Yahweh in the *Holy Bible*) is usually depicted in humanoid form.

Mohammedans. Islam rejects the characterization of god (Allah) either by portraiture or sculpture.

Australian aborigines. The Rainbow Snake god is depicted with the head of a kangaroo, the body of a python, and the tail of a crocodile; Altjira – the Sky God - as a large bird with the feet of an emu; and Adnoartine - the Lizard God - as its namesake, the lizard.

Conclusion

In his notebook, in or around the year 1837, Darwin made a sketch of an evolutionary tree, and in the 1859 edition of *Origin*, there appeared a more detailed 'Tree of Life' image. Had he been so minded, he might equally well have created a similar Tree, in order to depict how each of these various religions had evolved. The conclusion which is drawn from the above account is that:

1. Many different gods have appeared at many different times throughout human history.
2. Each god proclaims his own particular idiosyncratic message.

3. Each god is seemingly oblivious of, or hostile to, the presence of another.
4. Each god is absolutely convinced that his message is the correct one.
5. This, therefore, is a recipe for unending human conflict.

From whence did these disparate gods originate?

Could it be that man simply invented them? And if he did, then surely he may be forgiven. After all, how else could he make intelligible a world of earthquake and thunderstorm; tempest and drought; disease, famine, and conflict – events over which he had little or no control. Surely, then, it was only natural for him to long for a supernatural and omnipotent being who could offer himself and his family peace, security, and a good harvest each year; even the prospect of an afterlife? But what of the evidence of divine miracles, of god speaking to people or appearing to them in visions? Surely this cannot be lightly dismissed.

With the advances in psychiatry which have occurred in the twentieth/twenty-first centuries has come a realization that man is not a wholly rational being, and that the phenomena which he 'observes' or 'experiences' in the outside world, may, in fact, originate from within himself, in the form of delusions – idiosyncratic beliefs or impressions maintained despite being contradicted by reality or rational argument.[5]

The incidence in the population, as a whole, of those who experience delusions is estimated to be in the order of one person in every 30,000.[6] The cause of delusional disorder is unknown but it is sometimes associated with schizophrenia.

A hallucination is defined as an experience involving the apparent perception of something not present.[7] It may be something heard – an 'auditory hallucination', or something seen – a 'visual hallucination'. Hallucinations occur in severe disorders of mood, in organic disorders (relating to a bodily organ or organs) such as epilepsy, and in dissociative states (where a component of mental activity is split off to act as an independent part of mental life)[8] such as amnesia. 'Third-person' auditory hallucinations, where the person hears a voice which is making comments about him or her, is strongly suggestive of the presence of schizophrenia in that person.[9]

In addition, hallucinations may be induced by certain naturally occurring hallucinogenic substances including psilocin and psilocybin, found in the small genus of ('magic') mushrooms known as *psilocybe*; and tetrahydrocannabinol, the primary psychoactive component of cannabis and found in the plant of that name. Some modern-day drugs are also hallucinogenic, including lysergic acid diethylamine (LSD) and amphetamine. Therefore, a person is not necessarily telling a lie when he or she reports upon either hearing, seeing or believing in a god, if the vision or the voice upon which his beliefs are based has originated from within his or her own mind. Alternatively, it is possible that some who claim to witness 'divine revelations', do so either to create or to embellish a story, in the hope of achieving fame and high status amongst their followers, or for pecuniary gain.

It seems highly likely, in the light of this, that further examples of such man-manufactured gods may be expected to make their appearance in the future!

What were Darwin's views on the origin of religious belief?
In his book *The Descent of Man* Darwin states that:

> The belief in God has often been advanced as not only the greatest, but the most complete of all the distinctions between man and the lower animals. It is however impossible, as we have seen, to maintain that this belief is innate or instinctive in man. On the other hand a belief in all-pervading spiritual agencies seems to be universal; and apparently follows from a considerable advance in man's reason, and from a still greater advance in his faculties of imagination, curiosity and wonder. I am aware that the assumed instinctive belief in God has been used by many persons as an argument for His existence. But this is a rash argument, as we should thus be compelled to believe in the existence of many cruel and malignant spirits, only a little more powerful than man; for the belief in them is far more general than in a beneficent Deity. The idea of a universal and beneficent Creator does not seem to arise in the mind of man, until he has been elevated by long-continued culture.[10]
>
> ... until the faculties of imagination, curiosity, reason, &c., had been fairly well developed in the mind of man, his dreams would not have led him to believe in spirits [supernatural beings],[11] any more than in the case of a dog.[12]

In other words, man's belief in 'spiritual agencies' – i.e. 'gods' – only commenced when his mental faculties had become sufficiently well developed to permit it, which is tantamount to saying that god is exclusively a product of man's imagination.

Chapter 32

The Dinosaurs

Darwin was aware from the fossil evidence, including that which he himself had unearthed, that dinosaurs had once roamed the Earth, and he would have marvelled at the large number of new species which have been discovered since his time, together with fossilized dinosaur nests, eggs, faeces, footprints, and even imprints of their tails.

The period when these awe-inspiring and seemingly invincible creatures existed upon the Earth spanned virtually the whole of the Jurassic and all of the Cretaceous periods of geological time – i.e. from 205 million to 65 million years ago. However, whereas dinosaur fossils from these eras are to be found in abundance, such fossils are conspicuously absent from rocks of the succeeding Paleogene era (65 million to 56 million years ago). This indicates that dinosaurs became extinct at the end of the Cretaceous (from the Latin *creta* meaning 'chalk') period.

The so-called K-T Boundary is a geological stratum which marks the boundary between rocks of the Cretaceous and Paleogene eras. (The K derives from the German words *kreide*, meaning chalky and *zeit*, meaning period; and the T from Tertiary – the term used for the succeeding Paleogene and Neogene eras.) What comes as a surprise, is that the layer of clay or rock, just a few feet in thickness, which in many locations demarcates the K-T Boundary, is rich in iridium – an element which is rare on Earth, but commonly found in extraterrestrial bodies. The significance of this will be discussed shortly.

It was in 1677 that Robert Plot, Professor of Chemistry at Oxford University, published a description of a fossilized bone – believed to be part of the femur of a *Megalosaurus*, a large carnivorous bipedal dinosaur of the mid-Jurassic period[1], which he had discovered in Cornwall. This was the first dinosaur bone to be described in the literature. Subsequently, when the fossilized bones of many more similar types of creature were discovered, palaeontologist Professor Richard Owen (Darwin's former opponent in the 'Great Oxford Debate') recognized that these bones had certain distinctive features in common and classified them (in 1842) under the heading *dinosauria* (from the Greek, *deinos* – terrible and *sauros* – lizard).

Historically, it was from the archosaurs (ruling reptiles) that dinosaurs on the one hand and crocodilians (crocodiles, alligators, caymans) on the other were descended. Egg laying reptiles, dinosaurs tended to have long tails, long necks and small heads.

Their bodies were covered in scales composed of keratin, a durable, fibrous protein which provided, to varying degrees, waterproofing, and protection against predators. However, keratin is a poor insulator of the body when compared to fur and feathers, which is a point of great significance, as will shortly be seen.

The average size of a dinosaur approximated to that of a sheep. However, a full-grown specimen might range in size from the 15-inch-long *Parvicursor*, to the colossal, 120-foot-long *Argentinosaurus*, which weighed in at in excess of 100 tons. Dinosaurs differed from other reptiles in that their thigh bones (femurs) tended to point vertically downwards from the body (rather than to splay out laterally, like those of lizards), a feature which kept their bellies well off the ground. There were also other minor skeletal differences between the two types of creature. To date, approximately 1,000 different types of dinosaur have been identified (from their fossilized bones), and more are being discovered each year.

What caused the extinction of the dinosaurs?

Darwin puzzled over why dinosaurs, which included the largest and most powerful land animals ever to have lived, became extinct. Mere size and strength, as he pointed out in *Origin*, was no guarantor of survival.

Disease?

Some have argued that the dinosaurs were victims of a global pandemic caused by some unknown variety of virus or bacteria. However, at the time in question, which was the end of the Cretaceous era (demarcated by the so-called K-T Boundary), the single super-continent, known as Pangea, had already begun to split into several distinct entities, which makes a worldwide pandemic unlikely, because of the difficulties of transmission of the infective agent.

It is also commonly observed that if a species is exposed to an infectious disease, a certain portion of individuals belonging to that species are likely to develop an immunity to that disease. Myxomatosis did not destroy *all* the rabbits in Australia in the 1950s, and the great influenza pandemic of 1918–19 did not destroy *all* human life on Earth.

Climate too wet or too dry?

If it was the case that drought or flood had caused the extinction of the dinosaurs, then this would surely have also rendered extinct the mammals which co-existed with them.

Climate too hot or too cold?

In June 2011 the journal *Science* published an article by Robert A. Eagle (of the California Institute of Technology) and others, in which the authors described using 'clumped isotope thermometry to determine [lifetime] body temperatures from the fossilized teeth of large Jurassic sauropods [very large, quadrupedal, herbivorous dinosaurs with long necks and tails, small heads, and massive limbs].[2] The technique is based on the fact that the degree to which ^{13}C and ^{18}O (isotopes of carbon and oxygen respectively) tend to bond together is temperature related. Therefore, by measuring the extent to which these isotopes are found 'clumped together' in a dinosaur's bones and teeth (where they are deposited as these structures develop), the body temperature of the dinosaur can be ascertained.

The dinosaurs in question were *Brachiosauraus* (weighing in at between thirty-eight and fifty tons), and *Camarasaurus* (thirty-five to fifty tons). These creatures, so the authors discovered, maintained a body temperature of between 36°C to 38°C.[3] Until proved otherwise, it may therefore be assumed that all dinosaurs operated within this range of body temperature. The fact that dinosaurs were highly active, and could run for considerable distances at considerable speed – as is attested to by records of their fossilized footprints – also lends weight to the theory that these creatures were endothermic [dependent on, or capable of the internal generation of heat].[4]

The question is now asked – how does this optimal body-temperature range for dinosaurs compare with that of other creatures, for example, other species of reptiles, and mammals?

i. Reptiles of the non-dinosaur variety

In 1965, Bayard H. Brattstrom of the Department of Biological Sciences, California State University, published an article entitled 'Body Temperatures of Reptiles' in which he observed the following variations in the body temperature of snakes to be 9°C to 38°C; lizards 11°C to 46°C; US alligator 26°C to 37°C.[5]

ii. Mammals

The modern-day rat and shrew have a normal body temperature of between 38°C and 39°C, and it is likely (though not absolutely certain) that the same was true of the small varieties of mammal – shrew-like or rat-like creatures, that co-existed with the dinosaurs.

The question arises – would an unusual increase or decrease in temperature of the Earth's surface be responsible for annihilating the dinosaurs?

An inordinate and prolonged rise in temperature could undoubtedly have killed the dinosaurs. But if this was the case, both the mammals – whose fur would have made them even more vulnerable to overheating – and the non-dinosaur reptiles would have experienced a similar fate. On the other hand, what if there had been, at the time of the K-T Boundary, an inordinate and prolonged *fall* in temperature?

Mammals are insulated by a covering of hair or fur (short, fine, soft hair) – which is composed of keratin (the same durable, fibrous, protein material of which the scales of dinosaurs and other reptiles is composed). This acts by trapping pockets of air, and thereby enables heat to be retained in a similar manner to that by which double-glazed windows insulate a house. Mammals would therefore have been well able to withstand such cold conditions. So, too, would the non-dinosaur reptiles, which (for reasons described above) were able to operate at bodily temperatures so low that no dinosaur would have been able to survive them. It therefore comes as no surprise to find that, in such conditions, those mammals which had co-existed with the dinosaurs not only survived the Cretaceous era but thrived and diversified in the succeeding Paleogene era.

The problem for a dinosaur was that, whereas a mammal is cocooned within a furry, heat-retaining outer layer, this creature was more akin to a radiator, where heat continually and inexorably leached out from its body through keratinous scales which afforded virtually no insulation whatsoever. Paradoxically, some of the largest dinosaurs, such as the gigantic sauropods, were even more vulnerable than their smaller cousins, on account of their elongated necks and tails (the surface area/volume ratio of the neck and tail combined, exceeding that of the body by a factor of 3.5).

Furthermore, it has been suggested that dinosaurs, like their non-dinosaur reptilian 'cousins', had a heart and circulatory system (though this has not been proved). If so, this would have facilitated the rapid transfer of heat from the core to the periphery, making them even more susceptible to cold.

Were there any active steps which dinosaurs could have taken in order to keep warm? The answer is no, other than by the temporary expedient of involuntary shivering, or actively running around so as to generate heat within the musculature.

The consequences of hypothermia

In the phase of mild hypothermia the dinosaur's musculature would become unco-ordinated, resulting in slow and sluggish movements, and stumbling. And as hypothermia progressed and the creature's cellular metabolic processes began to fail, it would become confused and disorientated, death being the end result.

Is there evidence of a fall in the Earth's temperature at about the K-T Boundary?

There is scientific evidence that during the Cretaceous period, Earth experienced episodes of intense cold, and that one of these so-called cold events occurred during the *valanginian* era – which immediately preceded the K-T Boundary.[6] However, it is not known whether these 'cold events' were confined to the Arctic region, where the studies were conducted, or if they were worldwide phenomena?

What could have caused a drastic reduction in the Earth's temperature?

Asteroid collision

In 1977 US geologist Walter Alvarez was studying limestone rocks in the vicinity of Gubbio, a village of northern Italy. Here, sandwiched between rocks representing the end of the Cretaceous and those representing the beginning of the Tertiary layer, was a thin layer of red clay, about 2 to 2.5 centimetres in depth.

When Walter showed samples of this red clay to his father, physicist and inventor Professor Luis Alvarez of the University of California, Berkeley, the latter had the samples analyzed. They were found to contain the chemical element iridium, but in concentrations many times greater than that contained in the surrounding limestone – iridium being one of the rarest elements to be found in the Earth's crust. This led them to suggest that this iridium was present as the result of the collision of a huge asteroid with the Earth. Asteroids are bodies composed of rock, varying enormously in size, which orbit the Sun. Large numbers are found between the orbits of Mars and Jupiter, though some have more eccentric orbits.[7] Asteroids which originate from this so-called asteroid belt are rich in iridium.

The impact would have generated a massive fireball in which the asteroid would have been vaporized into a gas cloud, which would have reached high into the Earth's atmosphere. The theory is that this would have obscured the Sun, causing the temperature of the Earth to fall drastically. Some indication of the magnitude of the event is indicated by the discovery, at more than 100 locations worldwide, of similarly high concentrations of iridium in clay deposits laid down at the time of the K-T Boundary.[8]

The search for an asteroid crater, of the size demanded by such a colossal event and created at the relevant time, was now on.

The 'Chicxulub' Crater

A possible candidate was discovered in 1978 by geophysicist Glen Penfield, an employee of the Oil Company of Mexico, in that country's Yucutan Peninsula. The so called Chicxulub Crater (named after a nearby village) was 110-miles or so wide, and partially submerged beneath the ocean. The outcome was that, in September 1991, a scientific paper entitled 'Chicxulub Crater; a Possible Cretaceous/Tertiary Boundary Impact Crater on the Yucatan Peninsula, Mexico' was published in the journal *Geology*, by authors Alan Hildebrand, Penfield, and others. Having given their reasons for believing that the crater had indeed been created by the impact of a meteorite (defined as 'a meteor that survives its passage through the Earth's atmosphere such that part of it strikes the ground', they concluded that 'The age of the crater is not precisely known, but a K-T Boundary age is indicated'.[9]

There appeared to be a flaw in this argument, however. In March 2004 Swiss-born US palaeontologist Gerta Keller (Professor of Geosciences at Princeton University, New Jersey, USA) and others published in the *Proceedings of the Academy of Sciences of the United States of America* an article entitled, 'Chicxulub Impact predates the K-T Boundary Mass Extinction'. In it they declared that the aforementioned impact predated the K-T extinction by about 300,000 years.[10] This, however, was not the end of the story.

The 'Shiva' Crater

Mexico's Chicxulub Crater is dwarfed by the Shiva Crater, approximately 500 kilometres in diameter, which lies largely submerged off the coast of India, west of present-day Mumbai. It was discovered in 2009 by paleontologist Professor Sankar Chatterjee of the Museum of Texas Tech University, Texas, USA, and Dhiraj Kumar Rudra of the Geological Studies Unit, Indian Statistical Institute, Kolkala, India.[11] It was given the name, Shiva, after the Hindu god of destruction and renewal. But the unanswered question is, was this crater caused by meteoric impact, and if so, did the impact occur at the time of the K-T extinction?

Deccan volcanism

Another candidate for causing the Earth to cool was volcanic activity in the region of India's central Deccan Plateau which, according to Professor Gerta Keller and others, 'can now be positively linked to the K-T mass extinction'.[12]

Other factors

It may be that the Earth's cooling was unrelated to the activity of meteorites or volcanoes. For example, an alteration in the trajectory of the planet's elliptical orbit, which took it farther away from the Sun, might theoretically, produce the same effect.

Were the dinosaurs asphyxiated by poisonous gas?

Could it be that the alleged asteroid collision caused the planet to be enveloped in poisonous, life-extinguishing gases? If so, then why were mammals, birds, and reptiles not annihilated also?

Could the dinosaurs have starved to death?

Was it that the dinosaurs simply ran out of food, perhaps when Earth become covered in ash, following the alleged asteroid impact? This is an unlikely scenario, given the diversity of dinosaur species on the one hand, and the diversity of vegetation on the various continents, on the other. And again, if this was the case, how was it that other species survived?

* * *

Of these various theories, the likeliest one is that dinosaurs became extinct because, during one or more periods of global cooling, they were unable to maintain their body temperatures on account of heat leaching out from their poorly insulated skins. But is this the end of the story?

Chapter 33

Birds: The Only Surviving Dinosaurs

On 24 October 1867 Darwin's friend and colleague Thomas H. Huxley was examining, in Oxford University's Museum, the fossilized ilium (large broad bone forming the upper part of each half of the pelvis)[1] of a dinosaur. This, he observed, was avian-like. Furthermore, when the British Museum's specimen of an *Iguanadon* was reconstructed as a biped (rather than a quadruped), this led Huxley to remark that there was 'a considerable touch of a bird about the pelvis & legs'. The following year, at the Royal Institution, Huxley explained that the *Ratites* (a group of mainly large, flightless birds, such as the ostrich) were the descendants of Compsognathus-like dinosaurs.[2] (Compsognathus – a genus of small, bipedal, carnivorous theropod dinosaur.[3])

In other words, dinosaurs did not entirely become extinct, since at least one species survived, from which modern-day birds are descended.

Archaeopteryx

It was German palaeontologist Christian E. Hermann von Meyer who first discovered a fossilized feather belonging to this creature in 1860 at Solnhofen, Germany. The following year, when the first complete specimen was unearthed from the Solnhofen Plattenkalk (or Solnhofen limestone) Jurassic beds of Bavaria, he gave it the name *Archaeopteryx lithographica* (lithographic being the type of fine-grained, yellowish limestone in which the fossil was discovered). The fossil was sold to London's Natural History Museum where, as already mentioned, it was examined by Professor Owen. Several more such specimens were subsequently discovered in the same beds.

Archaeopteryx lived in the Late Jurassic period, 150 million to 145 million years ago; its name deriving from the Greek words *archaīos,* meaning 'ancient', and *ptéryx*, meaning 'feather' or 'wing'. Because of the exquisite preservation of the fossils, it has been shown that the skeleton of archaeopteryx is basically that of a small theropod dinosaur (a carnivorous dinosaur of a group whose members are typically bipedal and range from small and delicately built to very large).[4] However, attached to each arm was a feathered wing, and the body and tail were also feathered. At the end of each wing and foot were three separate fingers, each tipped with a curved claw.

Archaeopterix was, of course, long extinct by the time of the K-T Boundary and therefore it cannot be considered to be a direct ancestor of modern-day birds. However, it is considered to represent a transitional stage in the evolution of the theropod dinosaur to the biological class of vertebrates, Aves, that comprises the birds.

Since *Archaeopterix*, many more examples of feathered dinosaurs have been discovered. For example, in September 2009 an article entitled, 'Fossils Provide "Missing Link" between Dinosaur and Bird' by Richard Alleyne, Science Correspondent, was published by *The Telegraph* (online).

> Newly discovered fossils of a four-winged dinosaur have been hailed as the 'missing link' that finally proves [that] the prehistoric creatures evolved into birds. And five fossils recently unearthed in China show an eagle-sized Jurassic dinosaur that has feathers on its arms, legs, feet and tail. Paleontologists believe that the limbs would have developed into wings and enabled the tiny predators to take to the skies.
>
> The new feathered species were found in two separate rock formations in north east China – the tiaojishan, which would put the fossils at 168 million to 151 million years old, and the daohugou formation, which would make them 164 million to 158 million years old.

The dinosaurs in question were Troodontids, a predominantly small-bodied group of feathered theropods. One of them, which was given the name *Anchiornis Huxleyi* (Anchiornis – from the Greek *anchi*, meaning 'near to'; ornis – 'a bird', and Huxleyi – after Thomas Henry Huxley – Darwin's 'Bulldog'), was the size of a chicken, and had 'extensive plumage and profusely feathered feet'. Said Professor Xing Xu of the Institute of Vertebrate Palaeontology and Palaeoanthropology at the Chinese Academy of Sciences, Beijing, China,

> This fossil provides confirmation that the bird-dinosaur hypothesis is correct and supports the idea that birds descended from theropod dinosaurs, the group of predatory dinosaurs that include *allosaurus* and *velociraptor*.[5]

The dinosaur/bird precursor

The missing links between theropod dinosaurs and modern-day birds remain largely to be discovered. However, it is interesting to speculate as to what the dinosaur precursor of birds may have looked like, and to ask how did it survive the dinosaur extinction?

The dinosaur/bird which was extant at the time of the K-T Boundary extinction, and survived this catastrophic event, is likely to have been a relatively small creature, in which case its food requirements would have been equally so. It is also likely that it was compact in shape, with no excessively elongated extremities to make it vulnerable to heat loss.

Why did the 'bird-dinosaurs', alone of all the species of dinosaur, survive?

If it is the case that bird-dinosaurs were endothermic, and obliged, like modern-day birds, to maintain their body temperature at a comparatively high optimum level of between 40 to 42 degrees Centigrade – which was several degrees above that required for the remaining, non-avian dinosaur species – then the fact of their survival is remarkable. Furthermore, the fact that modern-day birds possess a heart and circulatory system, strongly implies that their fellow dinosaurs of the Cretaceous era were similarly endowed, making them even more susceptible to heat loss from the body. Why, therefore, did they not succumb to the alleged 'cold event' which allegedly occurred at around the time of the K-T Boundary?

The answer is that fully evolved feathers and down (fluffy plumage) act in the same way as hair and fur in preventing heat being lost from the body – i.e. by retarding movement of air across the surface of the skin and by trapping pockets of air within themselves. (In fact, feathers and down are made of exactly the same material as hair and fur – i.e. keratin). Furthermore, birds (like mammals) are able to increase the insulating capacity of their skins by 'puffing themselves up' – i.e. deliberately causing their fur or feathers to stand on end, thus increasing the depth of the insulating layer.

Chapter 34

The Eugenics Debate

'Eugenics', a term coined by Darwin's half-cousin Francis Galton, is defined as the science of improving a population by controlled breeding to increase the occurrence of desirable heritable characteristics.[1] Although many 'eugenicists' claimed to have been inspired by the work of Darwin, including Galton himself, the actual word only came into being after Darwin's death. On a more sinister note, some have blamed Darwin for producing ideas which set off a chain reaction, culminating in one of the greatest crimes against humanity in the history of the world – namely the Nazi Holocaust of the twentieth century. To what extent this is true will be discussed shortly.

Thomas Malthus

In 1798, in *An Essay on the Principle of Population*, Malthus declared that, by human intervention, 'The capacity of improvement in plants and animals, to a certain degree, no person can possibly doubt'. In other words, by means of selective cultivation from 'a wild plant', a 'beautiful garden flower may be produced', and similarly, by selective breeding, the most 'desirable qualities' of cattle or sheep. Therefore

> It does not … by any means seem impossible that by an attention to breed[ing], a certain degree of improvement, similar to that among animals, might take place among men.[2]
>
> The real perfectibility of man may be illustrated … by the perfectibility of plants. The object of the enterprising florist is, as I conceive, is to unite size, symmetry, and beauty of colour.[3]
>
> [However] As the human race … could not be improved in this way, without condemning all the bad specimens to celibacy, it is not probable that an attention to breed should ever become general …[4]

Malthus can therefore claim to be one of the very first eugenicists. He foresaw, however, that difficulties and dangers might befall those who attempted to apply the practice of eugenics to human beings, for

> an experiment with the human race is not like an experiment upon inanimate

> objects. The bursting of a flower may be a trifle. Another will soon succeed it. But the bursting of the bonds of society is such a separation of parts as cannot take place without giving the most acute pain to thousands: and a long time may elapse, and much misery may be endured, before the wound grows up [heals] again.[5]

Erasmus Darwin

Darwin's grandfather Erasmus, felt that it was important for a person contemplating marriage, to recognize in a potential spouse, the presence of inheritable diseases.

> As many families become gradually extinct by hereditary diseases, as by scrofula, consumption, epilepsy, mania, it is often hazardous to marry an heiress, as she is not infrequently the last of a deceased family [i.e. the remainder having succumbed to hereditary disease].[6]

(In fact, the first two diseases which Erasmus mentions above are of an infectious, rather than of a hereditary nature.)

Francis Galton

In 1883, the year after Darwin's death, Francis Galton's *Inquiries into Human Faculty and its Development* was published. In it, he developed more fully his ideas about 'eugenics' – a word which he himself invented and defined as

> the science of improving stock, which is not confined to questions of judicious mating, but which [gives] to the more suitable races or strains of blood a better chance of prevailing speedily over the less suitable than they otherwise would have had.[7]

In his 'Introduction', Galton states that 'We must free our minds of a great deal of prejudice before we can rightly judge of the direction in which different races need to be improved.'

> In every race of domesticated animals, and especially in the rapidly-changing race of man, there are elements, some ancestral and others the result of degeneration, that are of little or no value, or are positively harmful. We may, of course, be mistaken about some few of these … but, notwithstanding this possibility, we are justified in roundly asserting that the natural characteristics of every human race admit of large improvement in many directions easy to specify.[8]

Here, the words 'of every human race' are reassuring, in that Galton regards his aims and objectives as being of global application and relevance.

'Composite portraiture'

Under this heading, Galton describes how he took photographs of people who were engaged in various selected occupations. He then superimposed one image upon another in order to determine what traits, if any, the members of each group had in common with one another.[9]

An attempt to identify criminals, using the technique of composite portraiture

Galton was now able to contrast his 'composite portraits' of, say, the Royal Engineers, with 'the coarse and low types of face found among the criminal classes'.

> It is unhappily a fact that fairly distinct types of criminals breeding true to their kind have become established, and are one of the saddest disfigurements of modern civilization.

However, of the type of face which he regarded as being of a typically criminal nature, Galton was able to produce only two examples.[10] Despite the paucity of evidence he declared, 'I am sure that the method of composite portraiture opens a fertile field of research to ethnologists[11] – ethnology being the study of the characteristics of various peoples and the differences and relationships between them.[12]

The notion of an 'ideal race'

> It is the essential notion of a race that there should be some ideal, typical form from which the individuals may deviate in all directions, but about which they chiefly cluster, and towards which their descendants will continue to cluster.[13]

Galton now proceeds to describe those qualities which, in his view, are desirable and how they may be measured. They include stature, strength, capacity for labour, acuteness of hearing, and sense of vision and touch. Of particular value in an individual, is the ability to recall to mind visual images, number forms, conversations, and the written word.

Evolution by natural selection is a process which is neither intelligent nor compassionate

> The process of evolution on this earth, so far as we can judge, has been carried out neither with intelligence nor ruth [pity], but entirely through the routine of various sequences, commonly called 'laws', established or necessitated we know not how.[14]

A breeding programme designed to eliminate inherited 'barbarism'

> The hereditary taint due to the primeval barbarism of our [presumably the human] race ... will have to be bred out of it before our descendants can rise to the position of free members of an intelligent society ... .[15]

The dangers of 'overbreeding'
When it came to 'the stringent selection of the best specimens to rear and breed from ...' in order to promote 'the unlimited improvement of highly-bred animals ...', Galton issues a caveat. 'Overbred animals have little stamina ... an increasing delicacy of constitution ... [and] after a few generations ... fragility.'[16] He also points out that 'diminished fertility' is a feature of 'highly-bred animals'.

The limitations of 'selectivity'
Were a nation to 'banish a number of the humbler castes – the bakers, the bricklayers, and the smiths', said Galton, – then that nation 'would soon come to grief'. Therefore

> it will be easily understood that these difficulties, which are so formidable in the case of plants and animals, which we can mate as we please and destroy when we please, would make the maintenance of a highly-selected breed of men an impossibility.[17]

Nevertheless, this did not mean that the role of the eugenicist was redundant.

The favouring of a 'high race' over a 'low race': the concept of 'merciful eugenics'
A 'low race', said Galton, 'must be subjected to rigorous selection. The few best specimens of that race can alone be allowed to become parents, and not many of their descendants can be allowed to live'. Here he appears to be stating unequivocally that most of the offspring of 'low race' parents are to be eliminated. He continues:

> On the other hand, if a higher race be substituted for the low one, all this terrible misery disappears. The most merciful form of which I ventured to call 'eugenics' would consist in watching for the indications of superior strains or races, and in so favouring them that their progeny shall outnumber and gradually replace that of the old one.[18]

If a 'superior race' prevails, the outcome is enrichment, whereas if an 'inferior race' prevails, the outcome is unhappiness
'There exists a sentiment, for the most part quite unreasonable, against the gradual extinction of an inferior race.'

> That the members of an inferior class should dislike being elbowed out of the way is another matter; but it may be somewhat brutally argued that whenever two individuals struggle for a single place, one must yield, and that there will be no more unhappiness on the whole, if the inferior yield to the superior than conversely, whereas the world will be permanently enriched by the success of the superior.[19]

In the light of such remarks, Galton cannot complain if his remarks are construed as advocating the wholesale destruction of, what he terms, the 'inferior races'.

In Galton's view, the colonist may legitimately grab the spoils

It was likely, said Galton, that the Arabs would 'become one of the most effective of the colonizing nations … who may, as I trust, extrude hereafter the coarse and lazy negro from at least the [metalliferous] regions [those containing or producing metal][20] of tropical Africa'.[21]

The weakness of the Malthusian doctrine

> The check to over-population mainly advocated by Malthus is a prudential delay in the time of marriage … .

In other words, by postponing marriage to a later age, the female spouse's years for potential childbearing would be foreshortened. However

> such a doctrine … would only be followed by the prudent and self-denying; it would be neglected by the impulsive and self-seeking. Those whose race we especially want to have, would leave few descendants, while those whose race we especially want to be quit of, would crowd the vacant space with their progeny … .[22]

Malthus's proposals were therefore, in Galton's view, impracticable.

The way forward: celibacy for the 'less fitted'

> few would deserve better of their country than those who determine to live celibate lives, through a reasonable conviction that their issue would probably be less fitted than the generality to play their part as citizens.[23]

Galton and God

As far as the notion of God was concerned, it appears that Galton and Darwin were not far apart. For example, Galton refers to

> Our ignorance of the goal and purport of human life, and the mistrust we are apt to feel of the guidance of the spiritual sense, on account of its proved readiness to accept illusions as realities … .[24]

* * *

One of the problems with Galton's *Inquiries into Human Faculty and its Development* is that the author tends to vacillate in his opinions. For example, he appears, on the one hand, to propose extreme and drastic measures in order to achieve his eugenic goals, and on the other, to advocate a 'merciful' form of eugenics, based on voluntary participation.

* * *

At the International Health Exhibition held in London in 1884, Galton established an anthropometric laboratory (anthropometry being the scientific study of the measurements and proportions of the human body)[25] for the purpose of collecting statistics as to the acuteness of the senses, and the strength, height, and dimensions of large numbers of people.[26] In 1901 a quarterly journal entitled *Biometrika* was founded, to promote the study of biometrics (the statistical analysis of biological data) with Galton as its consulting editor. In 1904, The Galton Laboratory for Research into Human Genetics was established at University College London, where, in 1907, Galton initiated a scholarship in eugenic researches.

On 16 May 1904 Galton read a paper entitled 'Eugenics: its Definition, Scope, and Aims' to the Sociological Society at the School of Economics (London University), with mathematician and statistician Professor Karl Pearson FRS in the chair. In a 'highly selected society', ordered in accordance with his 'eugenic' philosophy, said Galton:

> The general tone of domestic, social, and political life would be higher. The race as a whole would be less foolish, less frivolous, less excitable, and politically more provident than now. We should be better fitted to fulfil our vast imperial opportunities [an indication that on this occasion, he was evidently referring to the British race].

As a way forward, Galton advocates the following courses of action:

> 1. Dissemination of a knowledge of the laws of heredity, so far as they are surely known, and promotion of their further study.
> 2. Historical inquiry into the rates with which the various classes of society

> (classified according to civic usefulness) have contributed to the [size of the overall] population at various times, in ancient and modern nations.
> 3. Systematic collection of facts showing the circumstances under which large and thriving families have most frequently originated; in other words, the conditions of eugenics. The definition of a thriving family … is one in which the children have gained distinctly superior positions to those who were their classmates in early life.

Galton advocates the compilation of a 'golden book' of such thriving families, to include details such as a person's 'race, profession, and residence; also of their own respective parentages, and of their brothers and sisters'.

> 4. Influences affecting marriage. If unsuitable marriages from the eugenic point of view were banned socially, or even regarded with the unreasonable disfavour which some attach to cousin-marriages, very few would be made.

Having advocated a system of eugenics which is based more on coercion than on voluntary participation, Galton insists that such a system

> must be introduced into the national conscience, like a new religion. What nature does blindly, slowly, and ruthlessly, man may do providently, quickly, and kindly. As it lies within his power, so it becomes his duty to work in that direction. The improvement of our stock seems to me one of the highest objects that we can reasonably attempt.[27]

Galton died on 17 January 1911. In his will he left the sum of approximately £45,000 for the establishment of the Galton Chair of Eugenics at the University of London with the wish, which was granted, that Karl Pearson be appointed as its first professor.[28]

The views of Alfred Russel Wallace on the subject of Eugenics

In 1890 an alternative view to Galton's was put forward by Wallace (who was to outlive Darwin by thirty-one years)

> I contributed to the *Fortnightly Review* an article on 'Human Selection', which is, I consider, though very short, the most important contribution I have made to the science of sociology and the cause of human progress. The article was written with two objects in view. The first and most important was to show that the various proposals of [writer and novelist] Grant Allen, Mr Francis Galton, and some American writers, to attempt the direct improvement of the human race by forms of artificial elimination and selection, are both unscientific and

> unnecessary; I also wished to show that the great bugbear of the opponents of social reform — too rapid increase of population — is entirely imaginary, and that the very same agencies which, under improved social conditions, will bring about a real and effective selection of the physically, mentally, and morally best, will also tend towards a diminution of the rate of increase of the population.
>
> A year later I contributed a paper to the Boston *Arena* ... and I pointed out that a more real and effective progress will only be made when the social environment is so greatly improved as to give women a real choice in marriage, and thus lead both to the elimination of the lower, and more rapid increase of the higher types of humanity.[29]
>
> I showed that the only method of advance for us, as for the lower animals, is in some form of natural selection, and that the only mode of natural selection that can act alike on physical, mental, and moral qualities will come into play under a social system which gives equal opportunities of culture, training, leisure, and happiness to every individual.[30]

In other words, according to Wallace, the goals of the eugenicists could just as easily be achieved by effecting an improvement in the social milieu.

Charles B. Davenport

In 1898, US biologist and eugenicist Charles P. Davenport became Director of the Biological Laboratory of the Brooklyn Institute of Arts and Sciences at Cold Springs Harbor, New York. Here, in 1910, he established the Eugenics Record Office, where a database containing records of the 'mentally deficient', the deaf, the blind, and the insane, were collected and stored. The outcome was that, between 1905 and 1972, an estimated 100,000 to 150,000 such people were sterilized, with or without their consent, as part of a federal government funded programme.[31]

In 1921 the Second International Eugenics Congress was held in New York City (the first having been held in London in 1912) and, the following year, the American Eugenics Society was established.

Karl Pearson

Whereas Francis Galton showed a degree of hesitation when it came to the question of eugenics in practice, Pearson was constrained by no such moral scruples. For example, in a lecture delivered to the Literary and Philosophical Society of Newcastle, UK in November 1900, he stated:

> If you have once realized the force of heredity, you will see in natural selection the choice of the physically and mentally fitter to be the parents of the next generation a most munificent provision for the progress of all forms of life.

Nurture and education may immensely aid the social machine, but they must be repeated generation by generation; they will not in themselves reduce the tendency to the production of bad stock. Conscious or unconscious selection can alone bring that about.

What I have said about bad stock seems to me hold for the lower races of man. How many centuries, how many thousands of years, have the Kaffir or the negro held large districts in Africa undisturbed by the white man? Yet their intertribal struggles have not yet produced a civilization in the least comparable with the Aryan [a word which, four decades later, was to have the gravest connotations, as will be seen]. Educate and nurture them as you will, I do not believe that you will succeed in modifying the stock. History shows me one way, and one way only, in which a high state of civilization has been produced, namely, the struggle of race with race, and the survival of the physically and mentally fitter race. If you want to know whether the lower races of man can evolve [into] a higher type, I fear the only course is to leave them to fight it out among themselves … .[32]

As regards the 'white man' who ventures to

lands of which the agricultural and mineral resources are not worked to the full … he could either settle down and live alongside the inferior race [or] the only healthy alternative is that he should go and completely drive out the inferior race.[33]

The struggle means suffering, intense suffering, while it is in progress; but that struggle and that suffering have been the stages by which the white man has reached his present stage of development.[34]

You will see that my view – and I think it may be called the scientific view of a nation – is that of an organized whole, kept up to a high pitch of internal efficiency by insuring that its numbers are substantially recruited from the better stocks, and kept up to a high pitch of external efficiency by contest, chiefly by way of war with inferior races, and with equal races by the struggle for trade-routes and for the sources of raw material and of food supply. This is the natural history view of mankind … .[35]

Mankind as a whole, like the individual man, advances through pain and suffering only. The path of progress is strewn with the wreck of nations; traces are everywhere to be seen of the hecatombs [sacrifices] of inferior races, and of victims who found not the narrow way to the greater perfection. Yet these dead peoples are, in very truth, the stepping-stones on which mankind has arisen to the higher intellectual and deeper emotional life of today.[36]

Pearson was, therefore, a eugenicist, a racist, a warmonger and a 'white supremacist'.

Chapter 35

Major Leonard Darwin

In 1912 the First International Eugenics Congress was held in London under the presidency of Darwin's son, Major Leonard Darwin. A soldier, politician and economist, Leonard was a former President of the Royal Geographical Society (1908–11), Chairman of the British Eugenics Society (1911–28) and its Honorary President from 1928 until his death in 1943.

In that February of 1912 Leonard delivered a lecture to the Cambridge University Eugenics Society entitled 'First Steps Towards Eugenic Reform'.

The aim of eugenics

> The primary object of Eugenics is, no doubt, to substitute for the slow and cruel methods of nature some more rational, humane, and rapid system of selection by which to ensure the continued progress of the race.

Potential drawbacks
Leonard was, however, aware

> that no reform is without some attendant evils, and that, in deciding how far any reform should be pushed, we have to attempt to ascertain the point at which the accompanying, evils would outweigh the advantages … . The possibility of inflicting a serious hardship on any individual, without any corresponding benefit in reality resulting from it to posterity, must, therefore, always be held in view.

Where coercion is necessary
There were certain situations, said Leonard, where 'marriages should be absolutely prohibited; as, for example, the marriage of two idiots'.

Where should the line be drawn?
Otherwise, said Leonard, 'Nature' offered no help whatever, when it came to deciding which individuals should be prevented from 'reproducing their kind'. And furthermore,

'her method of drawing the line is to kill off all who unaided are incapable of fighting the battle of life, a method which can no longer be tolerated'.

Why not breed humans in the same way as animals are bred?
Leonard quotes a breeder of prize-winning dogs who, having declared that 'I breeds a great many, and I kills a great many', posed the question, 'Then why not adopt this same successful method in dealing with mankind?' To which he (Leonard) replies

> that, although we do not doubt the possibility of making rapid progress by such means, yet we fully realize both the impossibility and the immorality of attempting to introduce the methods of the stud farm into human affairs.

Restrictions on marriage
Leonard concurred with what others had suggested previously, when he declared

> In considering the possibility of placing some check on the marriage of the less fit, it must be admitted that such a reform is likely to be first adopted amongst the most advanced races of mankind, and, moreover, that it would tend to produce a certain diminution in their numbers in comparison, that is, with the population which would have existed had no such reforms been introduced.

In other words, it was vital for the 'most advanced races' to keep up their numbers.

Immigration
Precautions were necessary, said Leonard

> against any gaps in our ranks being filled up by the immigration of less desirable stocks from abroad. If we are really intent on maintaining and improving the character of our race, we must, in fact, view with considerable suspicion our traditional policy of allowing nearly all comers to land on our shores.

The importance of defining 'ideal characteristics of the race', and the failure of Eugenicists to do so

> It has often been urged as an objection to all Eugenic practice that, before taking any steps intended to affect the characteristics of the race in the future, Eugenicists ought to decide on the ideal at which they are aiming, or on the exact type of man they wish to encourage. [However] on this point they, as a rule, are nearly silent.

What 'preventive legislation' should aim to achieve
'Our main efforts as regards preventive legislation' said Leonard,

> should be directed towards the reduction of the output of unquestionably undesirable types. In the present state of our scientific knowledge it would be as well to begin by endeavouring to make it impossible for those who are not only characterized by some signal heritable defect but who are also below the average both in bodily and in mental qualities, to reproduce their kind.

Potential criminals
As regards the 'large class' of people

> who are markedly inferior to the average of the nation if judged by many tests both mental and physical, [and] who have proved at an early age their incapacity to resist temptation, and who will therefore inevitably become criminals under existing social conditions, surely we ought at once to take precautions to ensure that the worst of them at all events should not, as it were, infect the coming generation with their defects … [Then] surely segregation for life with kindly treatment must in the interests of posterity be the fate of all who both fail in life in consequence of some signal heritable defect and have no redeeming qualities to compensate for such a defect.

A register of the 'abnormal'
Some system ought to be established, said Leonard,

> by which all children at school reported by their instructors to be specially stupid, all juvenile offenders awaiting trial, all ins-and-outs at workhouses, and all convicted prisoners should be examined by trained experts in mental defects in order to place on a register the names of all those thus ascertained to be definitely abnormal. In this examination both physical and psychological tests should if possible be included, in which case the reports thus obtained would afford a good foundation for selecting out the most unfit.

This alone, however, was not enough, for it was also necessary, 'from the Eugenic standpoint', to include in the register details of 'defects' in the relatives, including those who are

> unquestionably mentally abnormal, especially as regards … criminality, insanity, ill-health and pauperism … . If all this were done it can hardly be doubted that many strains would be discovered which no one could deny ought to be made to die out in the interest of the nation ….

Political difficulties attendant on 'eugenic reform': the need for education

> it is quite certain that no existing democratic government would go as far as we Eugenists think right in the direction of limiting the liberty of the subject for the sake of the racial qualities of future generations. It is here that we find the practical limitation to the possibility of immediate reform ….
>
> one of the first steps towards Eugenic reform must be the education of the public, an end to which our efforts should therefore now be directed.

Those who are severely disabled, either in mind or body

Given the political difficulties attendant on attempting to further the aims of the eugenicists, said Leonard,

> For the present we must content ourselves with dealing with the more obvious mental qualities. Passing on to physical defects, they should be regarded as being important in proportion to the amount of suffering they are likely to cause both to the individual actually afflicted and to his relatives and friends; those diseases producing permanently injurious effects necessarily, therefore, being ranked as the most serious.

Biometric screening

> When the family history of both parties [i.e. potential parents] is well known, the biometrician will doubtless before long be able to express in numbers the probability of the child of any marriage being afflicted with any heritable defect … .

By ascertaining whether the 'quality [i.e. disability] under consideration' is 'dominant in the Mendelian sense [i.e. according to the laws of inheritance devised by Austrian botanist Gregor J. Mendel]' or 'recessive', the attending physician

> will be in a far better position to give sound advice as to the probable results, and therefore as to the morality of any marriage, than one who merely trusts to his ordinary professional knowledge.[1]

In 1932, when the Third International Eugenics Congress was held at the American Museum of Natural History in New York City, the following article appeared in the *New York Herald Tribune* under the heading 'Eugenicists Hail Their Progress as Indicating Era of Supermen'.

> Major [Leonard] Darwin, who is the son of Charles Darwin, was unable to come to the congress because he is eighty-two years old. He presided at the first of these international gatherings and is still deeply interested in the progress of the work.
>
> His message was read last night by Dr R. A. Fisher [Sir Ronald Aylmer Fisher, evolutionary biologist, eugenicist, geneticist, and statistician] and was in part as follows
>
> My firm conviction is that if widespread Eugenic reforms are not adopted during the next hundred years or so, our Western Civilization is inevitably destined to such a slow and gradual decay as that which has been experienced in the past by every great ancient civilization. The size and the importance of the United States throws on you a special responsibility in your endeavours to safeguard the future of our race. Those who are attending your Congress will be aiding in this endeavour, and though you will gain no thanks from your own generation, posterity will, I believe, learn to realize the great debt it owes to all the workers in this field.[2]

Leonard's *The Need for Eugenic Reform*, was published in 1926 and

> dedicated to the memory of MY FATHER. For if I had not believed that he would have wished me to give such help as I could toward making his life's work of service to mankind, I should never have been led to write this book.[3]

Here, the significant phrase is 'if I had not believed that he would have wished', indicating an assumption on Leonard's part that Darwin would have endorsed his eugenic ideals and his proposals to put these ideals into practice, which is by no means certain.

Summary

Having observed how animals can be improved by selective breeding, eugenicists can hardly be blamed for wondering whether such methods might be applied to the human race also. However, most were aware of the moral and political dangers involved in 'playing God' – i.e. taking it upon themselves to decide who should live and who should die – all, that is, except Professor Karl Pearson, who appears to have possessed no moral conscience whatsoever.

Had Darwin been alive, what would have been his attitude to eugenics?

Darwin regarded himself as a scientist, first and last – i.e. a person who studies or has expert knowledge of one or more of the natural or physical sciences, 'science' being defined as the intellectual and practical activity encompassing the systematic study of the structure and behaviour of the physical and natural world through observation and experiment.[4]

Whether he would have approved of any or all, of his son Leonard's proposals in respect of eugenics in practice, is a matter of conjecture. The likelihood is, however, that he would have avoided the subject like the proverbial plague, as this statement, taken from his autobiography, indicates.

> I rejoice that I have avoided controversies, and this I owe to [Sir Charles] Lyell, who many years ago, in reference to my geological works, strongly advised me never to get entangled in a controversy, as it rarely did any good and caused a miserable loss of time and temper.[5]

There is, of course, a certain irony here, in that Darwin was involved in one of the greatest controversies of all time – that of evolution versus creation!

Chapter 36

Social Darwinism: The Deliberate Misrepresentation of Darwin's Ideas: The Nazi Holocaust

The term 'Darwinism' is defined, in present day terms, as

> The theory of the evolution of species by natural selection advanced by Charles Darwin [who] argued that since offspring tend to vary slightly from their parents, mutations [i.e. variations] that make an organism better adapted to its environment will be encouraged and developed by the pressures of natural selection, leading to the evolution of new species differing widely from one another and from their common ancestors.[1]

In her book *The Origins of Totalitarianism*, Hannah Arendt states that

> Darwinism was especially strengthened by the fact that it followed the path of the old might-right doctrine. Darwinism met with such overwhelming success because it provided, on the basis of inheritance, the ideological weapons for race as well as class rule ....[2]

It is true that Darwin, in *The Descent of Man*, did discuss the subject which later became known as eugenics. However, this he did in an objective but, above all, compassionate manner. Therefore, to use the term 'Darwinism' in the way that Arendt has done is to do the great exponent of the theory of evolution by natural selection a grave injustice for, although he observed that the 'fittest' – which usually meant 'the strongest' – tend to survive, never once did he subscribe to the 'might is right' philosophy, which means, in effect, that those who are powerful can do what they wish unchallenged, even if their action is, in fact, unjustified.[3]

What Arendt was, in fact, referring to above was what became known as 'Social Darwinism'

> the theory that individuals, groups and peoples are subject to the same Darwinian laws of natural selection as plants and animals. Now largely discredited, social Darwinism was advocated by Herbert Spencer and others in the late nineteenth and early twentieth centuries and was used to justify political conservatism, imperialism, and racism and to discourage intervention and reform.[4]

In other words, the works of Darwin were hijacked by extremists who misinterpreted them in order to serve their own ends.

Attempts to link Darwin with Hitler and the Nazi Holocaust

The Nazi Party was formed in Munich after the First World War. It advocated right-wing authoritarian, nationalist government and developed a racist ideology based on anti-Semitism and a belief in the superiority of 'Aryan' Germans.[5] In Nazi ideology, an 'Aryan' was defined as

> a person of Caucasian race not of Jewish descent. The idea that there was an 'Aryan' race corresponding to the parent Indo-European language was proposed by certain 19th century writers and was taken up by Hitler and other proponents of racist ideology, but it has been generally rejected by scholars.[6]

During the Holocaust an estimated 16 million people were murdered by Hitler's Nazi regime (between 1933 and 1945 when the Second World War ended). Of these, some 6 million were Jews, and this represented approximately 67 per cent of the Jewish population of Europe.

In his book *Hitler: Dictator or Puppet* the author has discussed at length the origin of Hitler's ideas, the sources of which were not Darwin but Doctor Leopold Poetsch and Arthur Schopenhauer, both of whom were racists and anti-Semites. Poetsch was Hitler's schoolteacher at the Realschule in Linz,[7] and Schopenhauer was a German philosopher who referred to the Jew as 'the great master of lies'.[8] However, the greatest influence of all on Hitler was Jörg Lanz von Liebenfels (1874–1954).

Liebenfels was an Austrian whose original name was Josef Adolf Lanz, who became editor, publisher and contributor to the anti-Semitic monthly journal *Ostara* (named after the ancient Germanic goddess of Spring). His writings are appropriately described as 'a potpourri of contemporary theories, most importantly of [Guido von] List's race theories [see below]'.[9] In his autobiography entitled *Mein Kampf* (*My Life*, published 1925–1926), Hitler, referring to his sojourn as a young man in Vienna, declares unashamedly in reference to *Ostara* that 'For the first time in my life, I bought myself some anti-Semitic pamphlets for a few pence.'[10]

Ostara was dedicated to the 'practical application of anthropological research for the purpose of preserving the European master race from destruction by the maintenance of racial purity'. It was lavishly illustrated with erotic pictures of blonde beauties being seduced by undesirable creatures, referred to by Liebenfels as 'beast-men' or 'ape-people'.

> Hitler possessed a great number of magazines, above all *Ostara* numbers. He was very interested in the content and also took the side of Lanz von Liebenfels very enthusiastically in discussions … .[11]

Liebenfels, in turn, had been influenced by Guido von List (1848–1919), a pan-Germanist (i.e. democratic, social-reformist, but anti-liberal and anti-Semitic). In 1911, List predicted war, prophesying that

> The Aryo-German-Austrian battleships shall once more … shoot sizzling from the giant guns of our dreadnoughts; our national armies shall once more storm southwards and westwards to smash the enemy and create order.[12]

List called for 'the ruthless subjection of non-Aryans to Aryan masters in a highly structured, hierarchical state'.[13]

Why, it may be asked, was Hitler so susceptible to the poisonous writings of those such as Liebenfels and List? In his book *Hitler: Dictator or Puppet*, the author has presented evidence that Hitler was, in fact, a sufferer from schizophrenia – a disorder which had a profound effect on his mind and way of thinking.

Houston Stewart Chamberlain

The final acknowledgment that Nazism in no way derived from 'Darwinism' – in the true sense of that word and as defined above – came from the English-born Nazi Houston Stewart Chamberlain (1855–1927).

Chamberlain was born in Southsea, Hampshire on 9 September 1855. His father William, was a commander in the Royal Navy (and, from 1874, Rear Admiral) and his mother Eliza, the daughter of a British naval captain.

At the age of fourteen, due to poor health, Chamberlain left England to visit health resorts on the continent, accompanied by his Prussian tutor Otto Kuntze, who extolled to him the virtues of Prussian militarism and introduced him to German history, literature and philosophy, including the works of artists and poets such as Beethoven, Schiller, Goethe, and Wagner. Kuntze remained his tutor for four years. At the University of Geneva, Chamberlain embarked upon a three-year study of various subjects, including philosophy, physics, chemistry and medicine. In Vienna in 1889, he began researching into plant physiology.

Another influence on Chamberlain's life was that of Joseph Arthur, Compte de Gobineau (1816–82) whose *Essai sur l'Inégalité des Races Humaines* made the case for the superiority of Nordics and Aryans – peoples whom he, Gobineau, forecast would decline, owing to their intermingling with other races.

In 1899 Chamberlain's own book *Die Grundlagen des neunzehnten Jahrunderts* (*Foundations of the Nineteenth Century*), was published in two volumes and, in it, he attributed the moral, cultural, scientific, and technological superiority of western civilization to the positive influence of the 'Germanic race' (which for him included Slavs and Celts). *Foundations* was greeted with rapture by the German Kaiser, Wilhelm II, who invited the author to his Court; thus began a lifelong friendship and correspondence between the two of them. 'It was God who sent your book to the German people, and you personally to me,' wrote Wilhelm to Chamberlain.

In 1915 Chamberlain, who regarded it as an act of treason by Britain that she had opposed Germany during the First World War, was awarded the Iron Cross for services to the German Empire; in 1916 he adopted German nationality.

In *Mein Kampf* Hitler refers admiringly to the 'principles of civil wisdom laid down by thinkers like Houston Stewart Chamberlain …'.[14] In 1925 the Nazi Party's official journal declared *Foundations* to be 'the gospel of the Volkish movement'. ('Völkische Bewegung' – a German, populist ethnic movement).[15] The book was also included as part of the curriculum of every German school and was to be found in all public libraries.

However, not only did Chamberlain *not* subscribe to Darwinism, in its true and original sense, he was openly contemptuous of it. For example, in his introduction to *Foundations*, he describes Darwinism as 'A manifestly unsound system'.[16] But his most acerbic comments on the subject are to be found in his biography of Immanuel Kant where, in his 'Author's Introduction', he speaks of 'A manifestly unsound system like that of Darwin … .' He then proceeds to make the following statements:

i. 'The recklessness with which Darwin frequently treats facts is beginning to be increasingly recognized.'[17]
ii. 'The Darwinian craze works such mischief ….'[18]
iii. 'Had Darwin, the incomparable observer of empirical phenomena, the man worthy of all honour, been in ever so slight a measure a thinker, he could not have failed to see that species in general is no direct natural phenomenon ….'[19]
iv. 'What a want of reflection disfigures the fundamental thoughts of Darwin and his followers.'[20]
v. 'Even Newton could have taught Darwin that empirical exact science never succeeds in making anything of the origin of natural phenomena …[21]'.

Chapter 37

Why Superstition May Be Preferable to Reason

Having discovered the theory of evolution by natural selection, Darwin lived to see it become increasingly accepted as a fact by scientists and others throughout the world. However, it was a source of great vexation to him a), that there remained those who refused, for one reason or another, to recognize its truth, and b), that millions preferred, instead, to hang on to what he regarded as superstitions – of which religion was a part, which all too often proclaimed messages that were diametrically opposed to his theory.

To those who have studied the writings and documentary films of ethnologist, evolutionary biologist and author Professor Richard Dawkins, it is clear that he is both bemused and perplexed by the failure, in many instances, of twenty-first century man to recognize and appreciate the truth of Darwin's theory. After all, this is not the Middle Ages, neither is it the nineteenth century, when the Great Oxford Debate – which ought to have laid the matter to rest once and for all – occurred. Said Dawkins:

> Rather than adapt to evidence, many of us, it seems, remain trapped in ways of thinking inherited from our primitive ancestors. Irrational belief, from dowsing [a technique for searching for underground water, minerals, or anything invisible, by observing the motion of a pointer (traditionally a forked stick, or paired bent wires) or the changes in direction of a pendulum, supposedly in response to unseen influences[1]] to psychic clairvoyance ['psychic' – 'relating to or denoting faculties or phenomena that are apparently inexplicable by natural laws': 'clairvoyance' – the supposed faculty of perceiving things or events in the future or beyond normal sensory contact[2]], has roots in early mankind's habit of attributing spirit and intention to natural phenomena such as water, the sun, a rock, or the sea. Even in the twenty-first century, despite all that science has revealed about the indifferent vastness of the universe, the human mind remains a wanton storyteller, creating intention in the randomness of reality.[3]

The use, by Dawkins, of the words 'inherited from our primitive ancestors', could be taken to mean 'passed on by text or word of mouth, down through the generations', or it could imply that there is a genetic basis for 'irrational belief'. However, he apparently believes that there is yet another explanation.

Dawkins's *meme hypothesis*

In order to explain how ideas, whether good or bad, spread from person to person and become lodged in their minds, Dawkins turned, not to the gene, but to the 'meme'.

> I think that a new kind of replicator has recently emerged on this very planet. It is staring us in the face. It is still in its infancy, still drifting clumsily about in its primeval soup, but already it is achieving evolutionary change at a rate that leaves the old gene panting far behind. The new soup is the soup of human culture. We need a name for the new replicator, a noun that conveys the idea of a unit of cultural transmission, or a unit of *imitation*.[4]

The name which Dawkins chose for this proposed new replicator was 'meme' - defined as an element of a culture or system of behaviour that may be considered to be passed from one individual to another by non-genetic means, especially imitation[5] (the word deriving from the Greek *mimema* – that which is imitated).[6]

> Examples of memes are tunes, ideas, catch-phrases, clothes fashions, ways of making pots or of building arches. Just as genes propagate themselves in the gene pool by leaping from body to body via sperms or eggs, so memes propagate themselves in the meme pool by leaping from brain to brain via a process which, in the broad sense, can be called imitation. If a scientist hears, or reads about, a good idea, he passes it on to his colleagues and students. He mentions it in his articles and his lectures. If the idea catches on, it can be said to propagate itself, spreading from brain to brain.

In *The Selfish Gene*, Dawkins quotes his colleague, psychologist Professor Nicholas K. Humphrey, as stating that

> Memes should be regarded as living structures, not just metaphorically but technically. When you plant a fertile meme in my mind you literally parasitize my brain, turning it into a vehicle for the meme's propagation in just the way that a virus may parasitize the genetic mechanism of a host cell. And this isn't just a way of talking — the meme for, say, 'belief in life after death' is actually realized physically, millions of times over, as a structure in the nervous systems of individual men the world over.[7]

Dawkins later elaborated thus:

> If memes in brains are analogous to genes they must be self-replicating brain structures, actual patterns of neuronal wiring-up that reconstitute themselves in one brain after another.[8]

> Consider the idea of God. We do not know how it arose in the meme pool. Probably it originated many times by independent 'mutation'. In any case, it is very old indeed. How does it replicate itself: By the spoken and written word, aided by great music and great art. Why does it have such high survival value? Remember that 'survival value' here does not mean value for a gene in a gene pool, but value for a meme in a meme pool. The question really means: what is it about the idea of a god that gives it its stability and penetrance in the cultural environment? The survival value of the god meme in the meme pool results from its great psychological appeal. It provides a superficially plausible answer to deep and troubling questions about existence. It suggests that injustices in this world may be rectified in the next. The 'everlasting arms' hold out a cushion against our own inadequacies which, like a doctor's placebo, is none the less effective for being imaginary. These are some of the reasons why the idea of God is copied so readily by successive mentions of individual brains. God exists, if only in the form of a meme with high survival value, or infective power, in the environment provided by human culture.[9]

No one would deny that one person can learn from another and, having done so, spread that person's ideas further afield. But Dawkins fails to explain why some people adopt an idea and believe in it, whereas others do not. An alternative explanation is, of course, that all superstitious beliefs, including those of a religious nature, are phenomena which are *entirely learned*.

For Dawkins to endorse the view that his hypothetical meme is an *actual physical structure* within the nervous system, is a bold step indeed, reminiscent of Darwin, and his (now disproved) theory of pangenesis. Nevertheless, said Dawkins,

> We have the power to defy the selfish genes of our birth and, if necessary, the selfish memes of our indoctrination. We are built as genes machines and cultured as meme machines, but we have the power to turn against our creators. We alone on earth, can rebel against the tyranny of selfish replicators.[10]

In other words, just as Dawkins regarded the gene as a selfish entity, so he appears to regard the meme in the same light. Subsequently, however, he issued this caveat.

> I had always felt uneasy [about] spelling this out aloud, because we know far less about brains than about our genes, and are therefore necessarily vague about what such a brain structure might actually be.[11]

Such conjectures on the part of Dawkins and Humphrey beg the question, what *is happening* inside the brain?

Chapter 38

The Ingrained Nature of False Beliefs

i. Curtailment of curiosity

By professing to have all the answers, religion may, in a believer's mind, obviate the need for further enquiry. Novelist, screenwriter and atheist Ian McEwan states that 'if you have a sacred text that tells you how the world began - the relationship between this sky god and you – it does curtail your curiosity'.[1]

ii. Brainwashing, especially of children.

To *brainwash* a person is to make him or her adopt radically different beliefs by using systematic and often forcible pressure.[2] Darwin warned against such tactics in his *Autobiography*, saying that we must not

> overlook the probability of the constant inculcation [the instilling of an attitude, idea, or habit, by persistent instruction][3] in a belief in God on the minds of children producing so strong and perhaps an inherited effect on their brains not yet fully developed, that it would become as difficult for them to throw off their belief in God as for a monkey to throw off its instinctive fear and hatred of a snake.[4]

(This sentence was subsequently removed from the text of Darwin's *Autobiography* at the request of his wife Emma, who made her request three years after her husband's death.)

The Jesuits, or the Society of Jesus, is a Roman Catholic order of priests founded by St Ignatius Loyola, St Francis Xavier and others in 1534 to do missionary work. The order was zealous in opposing the Reformation.[5] The maxim of the Jesuits is, 'Give me a child for his first seven years and I'll give you the man'. In other words, the Jesuits appreciate the fact that young minds are particularly vulnerable to religious indoctrination, enforced by the constant exposure to 'sacred texts', music, iconography, etc. Similarly, it is no accident that some other religions also have their faith schools.

iii. Ignorant, biased, or politically correct teachers

Dawkins was appalled to discover that a teacher of science in an English school believed, despite overwhelming evidence including that provided by the science of carbon dating, that the Earth was less than 10,000 years old. He was equally surprised that some schoolteachers were unwilling to challenge their pupils' beliefs, in circumstances where the holy scriptures said one thing and science another.

iv. The stifling effects of religious or political ideology: the censoring of information

Since Darwin there has been an unending succession of assaults against intellectuals, dissidents and freethinkers, amongst the most notorious being the Chinese so-called Cultural Revolution that took place in between 1966 and 1976. Initiated by Mao Zedong, Chairman of the Communist Party of People's Republic of China, its stated aim was to enforce socialism, remove capitalist, traditional and cultural elements from Chinese society and impose Maoist orthodoxy. Today, such persecution continues the world over, as Amnesty International – a worldwide, non-governmental organization, whose objective is 'to conduct research and generate action to prevent and end grave abuses of human rights, and to demand justice for those whose rights have been violated' – will readily confirm.

Another tool in the hand of the modern-day despot is censorship, which might mean jamming satellite radio and television broadcasts from abroad; ordering material to be removed from the internet; or banning foreign newspapers, lest their ideas contaminate the minds of the despot's subjects.

v. Peer pressure

This may also be a strong influence in curtailing freedom of thought.

vi. Threats to the unbeliever, of violence and intimidation in this world, and eternal damnation in the next

vii. Genes and belief

Simon Baron-Cohen, Professor of Developmental Psychopathology at the University of Cambridge, points out that

> If a trait or behaviour is even partly genetic [i.e. inherited], we should see its signature showing up in twins. If the trait or behaviour in question does not differ much between MZ and DZ twins, then one is forced to conclude that genes play little if any role in the behaviour. This is because MZ twins are like genetic clones (they are genetically identical, so share 100 per cent of their genes), while DZ twins are genetically no different to any pair of siblings (they share on average 50 per cent of their genes).[6]

(MZ - Monozygotic – derived from a single ovum, and so identical. DZ - Dizygotic – derived from two separate ova, and so not identical.)[7]

Research having demonstrated that there is a greater similarity in the religious behaviour of MZ twins, as compared with DZ twins, this is taken to mean that there is a genetic basis for religiousness. Shortly, it will be demonstrated that this notion is not as farfetched as it at first appears.

viii. Hormones and belief

It has been found that during meditation, of which prayer is a part, there is a rise in the blood levels of the neurotransmitters dopamine and serotonin.[8] Could it be that the feel-good factor which these substances provide has an addictive property which induces the individual concerned to become more religious?

Chapter 39

Genetic Science Vindicates Darwin and Provides an Explanation for Variation

Darwin would have been gratified to know that the science of genetics not only corroborates the truth of his theory of evolution, but also provides infinitely greater possibilities for understanding the precise manner in which organisms are genetically related to one another, and sheds light on how they have evolved.

The cell is the smallest structural and functional unit of an organism, and typically consists of cytoplasm and a nucleus enclosed in a membrane. (However, not all cells contain nuclei, for example, bacteria and human red blood cells). Cells contain threadlike structures called chromosomes, the number of which varies according to species. Human cells, for example, of which there are billions, contain forty-four chromosomes, arranged in pairs (each pair consisting of one chromosome inherited from the father and one from the mother), plus a pair of 'sex chromosomes' (two X chromosomes in females, one X and one Y in males). Between them the chromosomes contain the entire information – blueprint – for the organism, and also a mechanism whereby that organism can reproduce. This is explained as follows.

Chromosomes contain deoxyribonucleic acid (DNA) in the form of a long, double-stranded molecule in the shape of a double helix, which has been likened to a ladder, twisted into a helical shape. The total length of DNA in each cell is approximately one metre, and the ability of the two DNA strands to separate enables new bases to attach to each, and hence another double strand to form – a process which occurs in cell division. The fact that DNA is to be found in the nuclei of all the cells of all living organisms is in accordance with all forms of life being interrelated and having evolved from a common ancestor.

DNA has a backbone structure of sugar and phosphate molecules, to which are attached nitrogen-containing bases, four in number – adenine, cytosine, guanine and thymine. It is the bonds between these pairs which holds the double helix together, cytosine always bonding to guanine, and adenine to thymine. Each human being has in excess of three billion of these so-called base-pairs.

A gene, of which each human being possesses about 20,000-25,000, is a length of chromosome containing a distinct sequence of nucleotides, and upon the order of this sequence – or code – depends what type of protein is produced. The gene achieves this by forming ribonucleic acid (RNA) which transfers the coded message from the nuclear DNA to the cell's cytoplasm, where amino acids are pieced together accordingly. In a given cell, the vast majority of genes are switched off, and only those which relate to the function of that particular cell are operational.

Different genes code for different proteins, which are the structural materials from which the cell is built, and for enzymes (which are also proteins), which regulate the chemical reactions which occur within the cell, and therefore its activity. Although some traits are coded for predominantly by a single gene, for the overwhelming majority of traits, the participation of a number of genes is required.

Genes are located in pairs, opposite to one another on the chromosome pair - one inherited from the father (as is the chromosome) and one from the mother (likewise). The two genes – which are called alleles – may be identical, or one may be a variant of the other, in which case only the so-called dominant gene is expressed (and not its recessive partner).

The Human Genome Project

A genome is defined as the complete set of genes or genetic material present in a cell or organism.[1] The Human Genome Project was begun in 1989 with the aim of identifying all of the approximately 20,000 to 25,000 genes contained within human DNA, and determining the sequences of its 3 billion base-pairs. This was successfully achieved fourteen years later, in 2003.

The genes of animals and plants

The cells of every animal and every plant contain similar quantities of DNA, though the number of chromosomes and genes which they contain varies greatly. For example:

	Chromosomes	Genes
Mouse	25,000	40
Fruit fly	13,600	10
Arabidopsis thaliana (a member of of the mustard family of plants)	25,500	5
E. coli (a bacterium)	3,200	1

However, how similar or different one organism is to another is dependent upon how much of their respective DNA sequence – or 'code' – they share in common. For example

> We humans share 99 per cent of the same DNA sequence as chimpanzees, from whom we split 6 million years ago, 90 per cent with mice (100 million years), and even 31 per cent with yeast (1.5 billion years).[2]

Not for nothing has man sometimes been referred to as the 'fifth ape'!

A DNA-based evolutionary tree of life

The discovery of DNA and the mapping of the genomes of various animals and plants will one day enable a comprehensive DNA-based tree of life to be constructed.

Variation explained

All his working life Darwin grappled, unsuccessfully, with the question of how variation in nature comes about. Was it something that was determined a), by chance b), by the external environment or c), at the volition of the individual?

It was Austrian priest and botanist Gregor J. Mendel (1822–84), who had provided an early explanation of variation. In the garden of the Augustinian Abbey of St Thomas in Brno (then the capital city of Moravia, now the Czech Republic), of which he became abbot, Mendel conducted experiments in order to discover how the characteristic features of plants were inherited. First, he obtained tall plants which bred true (i.e. produced only tall plants when self-pollinated or cross-pollinated with other tall plants) and short plants which bred true. When he crossed true-breeding tall plants with true-breeding short plants, he discovered that all the offspring produced were tall. He concluded that plants receive one character from each of its parents, tallness being a dominant characteristic and shortness being a recessive, or hidden, characteristic, which only reappeared in subsequent generations. But the answer as to how variation operated on a molecular level, would only come through advances in science, and this would not be until long after Darwin's death.

Asexual reproduction (mitosis)

Asexual reproduction involves 'a type of cell division that results in the formation of two "daughter cells", each having the same number and kind of chromosomes as the parent cell'.[3] It is the primary form of reproduction for single-celled organisms, such as bacteria, and also for many plants and fungi. Here, variation may occur as the result of mutations brought about when mistakes happen in the course of the cell copying its DNA in preparation for cell division.

Sexual reproduction ('meiosis')

Sexual reproduction is the primary method of reproduction for the vast majority of animals and plants. In the case of human beings, when gametes (spermatozoa and ova) are created, this involves the production of four (diploid) cells containing only twenty-three single – i.e. unpaired – chromosomes from a standard (haploid) parent cell containing twenty-three pairs of chromosomes. In this process, there is an exchange of genes between those derived from the father and those derived from the mother. The newly produced chromosomes, and therefore the four daughter cells are each unique, being genetically different from one other, and from the parent cell. Furthermore, when two gametes (sperm and ovum) come together to form a new diploid cell (or zygote, which then undergoes cell division to create the embryo) there is a further exchange of genes (alleles) between the two newly-formed chromosome pairs.

Organisms which reproduce by meiosis therefore contain an enormous mixture of genes, deriving not only from their immediate parents, but also from their ancestors, in the ways described above. From this, it can be seen not only that variation is an inevitable consequence of sexual reproduction, but that the possibilities for such variation are endless.

Gender variation is itself an example of variation. If an egg (ovum) is fertilized by a sperm containing a Y chromosome, the resulting zygote will be XY or male. On the other hand, if an egg is fertilized by a sperm containing an X chromosome, the resulting zygote will be XX – or female.

But does the above account provide the *only* explanation for variation?

A newly discovered, epigenetic mechanism for variation

Tim Spector, Professor of Genetic Epidemiology at King's College, London has drawn attention to the fact that there is another mechanism by which variation may arise. He points to the work of Pilar Cubas and others of the John Innes Centre for Plant Science and Microbiology, Norwich, UK who discovered an alternative variety of the wild flower common toadflax. This variation, however, was not the result of a change in the structure of the plant's DNA (as is the case with mutations), but to a phenomenon called methylation. In the toadflax variation

> a key gene (called Lcyc) is extensively methylated and in the normal plant it is not.
>
> What methylation means is that at certain sites (usually cytosine bases) of the gene's DNA, small chemical methyl groups floating around the cell attach themselves to it … . This has the effect of stopping the gene producing a protein.

In other words, methylation stops the gene from functioning, by switching it off and thereby preventing it from expressing itself. It was also shown that the methylated gene could be passed on to subsequent generations.[4]

Spector points to *aridopsis*, a small, flowering plant related to the cabbage, as an example of how a gene may be switched off by this mechanism.

> In response to prolonged cold (as in winter), the Flowering Locus C gene which normally prevents flowering is methylated and deactivated, allowing this variety to flower in the spring. This trait is then passed on to the next generation [when the plant will once again flower in the spring], even if there is no cold winter.[5]

Although the example of a plant has been cited above, it is likely that the phenomenon of methylation of a gene or genes, and the inheritance of such epigenetically altered genes (epigenetic – resulting from external rather than genetic influences),[6] together with the characteristics that they code for, is to be found in all forms of animal and plant life. Spector cites several examples of the role which epigenetics may play in various aspects of life:

i. **Mindset** (the established set of attitudes held by someone)[7] and personality

> we now know that both genes and their related mindsets, which we thought to be hard-wired, can be modified and reset along with the traits and personalities that define us as individuals.[8]

ii. Belief systems
There is a suggestion, says Spector, that 'even our patterns of beliefs could be altered epigenetically – with faith genes switched on and off.'[9]

iii. Maltreatment in infancy
Studies involving rodents, says Spector

> show that maltreatment in the first week of life causes certain genes to be switched off epigenetically. The glucocorticoid receptor gene controlling the stress response is the best example, where a methyl group is attached and so the gene cannot be expressed. This leads to a cascade of changes in many other genes related to emotion and stress, and can last a lifetime.

In respect of human beings, when previously maltreated children become parents

> because of their abnormal methylation [they] often fail to bond with their own children because their empathy or bonding genes are not working normally. This leads to an increasing cycle of dysfunctional parenting or abuse.

Finally

> this methylated gene will be present in the sperm or egg of the child [i.e. who was once maltreated, but is now an adult] and then usually passed on to the next generation when they reproduce.[10]

iv. Mother/child bonding
Similarly, 'dangerous alterations in gene methylation' may occur when the 'bond of interaction' between mother and child 'goes wrong'.[11]

v. Stress
Spector also suggests that epigenetics can 'explain how stress translates into health problems' as a result of methylation 'epigenetically deactivating' the 'immune genes [i.e. the genes which code for immunity]', leading to a weakening of the immune system in those who are experiencing stress.[12]

vi. Cancer
In cases of cancer, says Spector, it is generally the case that the DNA of cancer cells is

> under-methylated – enabling many genes that are normally suppressed literally to run wild. [On the other hand] a few DNA areas show the opposite: they are hyper-methylated and the genes suppressed. These genes are the body's built-in protection system; the tumour-suppressor genes that keep the DNA under control.[13]

vii. Pain
It is also the case that certain 'pain genes' can be switched on or off epigenetically.[14] (The difference in the degree to which this occurs may explain why some people's pain threshold is higher or lower than others.)

Because methylated genes may be passed on down the generations, says Spector, both the environment and the behaviour of our 'parents and grandparents may influence us in a number of ways – modifying our growth, altering our brain development, and affecting our risk of diabetes and heart disease'.[15] Therefore, says Spector, in respect of the

> inheritance of acquired characteristics' [or so-called] 'soft inheritance', proposed by Lamarck and accepted as possible by Darwin … despite the ridicule this received for most of the last 150 years, we now know that it can occur.[16]

Spector also points out that the inheritance of characteristics acquired in an epigenetic manner

> is the parallel, faster route by which we human beings adapt to our surroundings, and also explains many of the emerging ideas of how we are moulded into individuals.[17]

In other words, variation which is brought about epigenetically is more advantageous to the organism concerned, and occurs at a more rapid rate than would otherwise be the case.

> The most important lesson that we've learnt is that you can change your genes, your destiny and that of your children and grandchildren. It really does matter what you do to your body, and importantly what your grandparents did to theirs many years ago.[18]

The discovery of this new, epigenetically driven mechanism for variation (which in no way invalidates Darwin's theory of evolution by natural selection) is exciting because it opens up an entirely new field of research into how diseases – both physical and mental – are caused, and into possible therapies.

Chapter 40

Darwin and Downe's Church of St Mary the Virgin

When the Darwins arrived at Downe in the autumn of 1842 the vicar of its Parish Church of St Mary the Virgin was the Reverend James Drummond.

> There was no Unitarian chapel in the vicinity, and the family attended the local Anglican church, St Mary's, each Sunday. All the children were baptized and confirmed in the Church of England. The whole family took the sacrament, although Emma [mindful of her Unitarian beliefs] used to make the children turn around and face the back on occasions when the rest of the congregation recited the Athanasian Creed. Around 1850, Darwin himself stopped attending services. He would accompany the family to church and would often wait outside, chatting with the village constable, or would stroll around the village until the service ended. Yet despite absenting himself from worship, Darwin was actively engaged in church affairs. He took a lead in local charities, supervised church and school finances, and worked to uphold the status of the church in the community.[1]

Darwin was a man of principle, and to attend church would be to betray his principles, to be a hypocrite. So why did he support the Church in other ways? Perhaps for sentimental reasons, and to keep on good terms with the vicar and his flock, but primarily to propitiate his wife Emma.

In assuming what he saw as his 'pastoral responsibilities in the village ... Darwin worked closely with the Anglican incumbent John Brodie Innes [Drummond's successor], who became Perpetual Curate of Downe in 1846'. This, despite the fact that Innes, 'A Tory and High Churchman ... [High Church – of, or adhering to, a tradition within the Anglican Church emphasizing ritual, priestly authority, sacraments, and historical continuity with Catholic Christianity[2]] preached eternal torment ...' – i.e. eternal punishment in Hell for sinners and unbelievers; a doctrine Darwin had previously described as 'damnable'.

> But their political and doctrinal differences were glossed over, it would seem, in their shared sense of duty toward the community, especially its poor. They collaborated on the running of village charities, a Coal and Clothing Club [a local charitable institution which supplied parishioners with coal and clothes in exchange for regular contributions], and a Friendly [local savings and insurance] Club. Darwin served as treasurer of both organizations, and would read out the accounts to members, who assembled on his lawn for regular meetings.[3]

'To the end of his life, Innes remained unpersuaded by Darwin's theory of evolution.' Nonetheless, the clergyman, who shared with Darwin an interest in natural history, was 'always keen to pass on useful facts of natural history … [and, for example] reported on toads found in the stone cuttings of the new railway'.[4] In fact, Innes described Darwin as 'one of my very most valued and dearest friends'.[5]

On 15 December 1861 Darwin told Innes that he would be delighted to adopt, at Innes's request, the latter's son Johnny's dog. 'Quiz' would be 'taken great care of, & never parted with & when old & infirm shall pass from this life easily. Most truly yours C. Darwin'.[6] Sadly, however, the following May, Darwin informed Innes:

> I have bad news about Quiz: perhaps you had better not tell your son for a time. He has been killed; it was done instantaneously by a gun. We were forced to do this, for he would fly at poor people, & one day bit a child & two days after a beggar woman & we had an awful row about it.

However, another reason was that

> we could not stop him having fearful battle with Tartar [Darwin's own dog]; I had such a job one day in separating them both streaming with blood; & this was *incessantly* happening. Poor little Quiz had, also, got so asthmatic that he could not run, so that altogether we had no choice left us, though we were very sorry about it.[7]

In the year 1862 Innes relocated to Scotland, to an estate which he had inherited. Nevertheless, he

> retained the advowson of the village [the right to recommend or to appoint a suitable Anglican clergyman to the vacant living i.e. position as vicar],[8] and was thus responsible for the appointment of a resident curate and the maintenance of a local parsonage. The village of Downe did not fare well under this system. The living was comparatively small, and the local parsonage had been sold. In fact, some years before the Darwins arrived, the parsonage had actually been Down House. The details are not known, but Innes evidently

had property of his own in the village, and did not need a parsonage. When he left for Scotland, he tried to secure the purchase of land for a new parsonage, but was unsuccessful.

Beginning in 1867, Downe's parishioners experienced troubles securing the services of a reliable replacement for Innes. Darwin complained of the prolonged absence of the curate at the time, Samuel James O'Hara Horsman, and reported that, owing to difficulties in accessing church funds under Horsman's care, he had had to advance the salary of the schoolmaster. Darwin also acted as intermediary for Horsman, who excused his long absence as due to his needing a 'change of air' and being invited by some friends, 'who kindly took me about in their yacht & otherwise made it pleasant to me'.

In addition, said Horsman, he was induced to stay away because of 'the wretched & miserable lodgings at Downe' and all kinds of 'wicked reports & misrepresentations about me'.[9]

Darwin wrote Innes on 20 January 1868 to say,

> I am much obliged to you for sending me your Sermon … . You would have been pleased & w[d] have admired our church [at Downe] this Christmas as it was ornamented with great taste.[10]

Darwin expressed concern to Innes about the Reverend Horsman in June:

> I was sure that you wd feel much annoyment with respect to Mr Horsman's conduct in your parish. On June 2 Mr H. wrote to me a foolish letter, in which he said he believed that he held some balance on the [national] school account.

National schools, such as the one at Downe, provided an elementary education to the children of the poor, in accordance with the teaching of the Church of England.

> I wrote immediately in answer asking him to send me a cheque for the amount together with all accounts & documents relating to the school. I told him that on their receipt I wd send him a formal acknowledgment; but I have not heard *a word from* him since. The accounts of the Nat. school were audited up to Dec 31. 1867, & Mr H. has a balance in hand of £8—4—10 … . [Furthermore] the schoolmaster was not paid for last quarter, & now a second qr will soon be due, both of which I will advance.
>
> He [Horsman] owes, I am told, some few bills in the village. There is one more serious matter; he was curiously anxious to get up a subscription for the new organ & some of the Lubbocks [of High Elm, Downe] have suggested that he may have pocketed the money & never paid for it.

> Mr H. almost entirely neglected the school & considerable repairs were found necessary so that I fear the school account will be in a bad way. [Darwin ended his letter] I almost think he is more an utter fool than knave.[11]

Horsman's replacement was John Warburton Robinson, who proved to be no better, for, on 1 December, Darwin informed Innes that the new curate

> has suddenly left us to stay for 3 months in Ireland, & as I did not anticipate anything of the kind, I passed over the school account to him … . The curate, whom Mr Robinson has sent here does not appear any great [acquisition]. Mr Horsman, now that he is known to have been a complete & [premeditated] swindler (for no other interpretation as it seems to me can be put on his conduct about the Organ) has done much injury in the Parish & some of the subscribers to the School were actually afraid to pay the subscriptions to Mr Robinson apparently merely for [i.e. by reason of his] being a clergyman; & what they will think now that he has gone off for 3 months, I know not …
>
> As I fully believe that you are anxious to do all the good that you can to your parish, I am sure you will allow me to say that unless you can very soon make some fixed arrangement, so that some respectable man may hold the living permanently, great injury will be done here, which it will take years to repair, & what you will consider of importance, the Church, will be lowered in the estimation of the whole neighbourhood. Already so staunch a tory [a member or supporter of the Conservative Party, which traditionally supported the Church of England] & [church-goer], as old Mr Abraham Smith goes to dissenting chapel & [propounded] the doctrine so astounding as coming from him, that perhaps the disestablishment of the English Church wd be no bad thing.[12]

Exasperated as he was by these events, Darwin was anxious that his local church should not be neglected by its clergy. However, he evidently found the 'dissenting' doctrine more to his taste than the Anglican one!

Nine days later, Darwin wrote to inform Innes that

> rumours are very common in our village about Mr Robinson walking with girls at night. I did not mention them before, because I had not even moderately good authority; but my wife found Mrs Allen very indignant about Mr R's conduct with one of her maids.
>
> I do not believe that there is any evidence of actual immorality. As I repeat only second hand my name must not be mentioned. Our maids tell my wife that they do not believe that hardly anyone will go to Church now that Mr R. has returned.
>
> What a plague this Parish does give you.[13]

Darwin might as well have added, after this last remark, 'and me also'!

Following the scandals of Horsman and Robinson, Darwin

> now kept the books of the Sunday School and the National School as well, and took personal charge of raising money for the upkeep of church buildings, as well as overseeing the repairs. 'The Church will be lowered in the estimation of the whole neighbourhood', Darwin warned.[14]

In 1869 yet another clergyman, the Reverend Henry Powell, was appointed Vicar of Downe. A relieved Darwin described him as

> a thoroughly good man & gentleman. Does good work of all kinds in the Parish, but preaches, I hear, very dull sermons.[15]

Emma told writer and philanthropist Frances P. Cobbe in February,

> I think the course of all modern thought is 'desolating' as removing God Further off. But I do not know whether his [Darwin's] views on the moral sense would exclude Spiritual influence though not included in his theory – So you see I am a traitor in the camp.[16]

In other words, Emma is hoping against hope that her husband has not totally abandoned his Christian faith.

In that year Powell was succeeded by George Sketchley Ffinden, who served as Vicar of Downe for the next forty years. Like Innes, Ffinden

> was a Tory High Churchman, but unlike his predecessor he was unwilling to share parish leadership with the likes of the Darwins. A series of conflicts ensued, leading quickly to an impasse. Emma Darwin's regular use of the village schoolroom as an evening reading room for workers was opposed by Ffinden on grounds that the space, so used, would not be left in salutary condition for scholars. Darwin appealed directly to the School Committee, prompting this curt reply from the curate: 'As I am the only recognized correspondent [person authorized to communicate with the authorities] of the School according to rule 15, Code 1871, I deem such a proceeding quite out of order.'[17]
>
> Darwin's clash with Ffinden might be attributed to ideology. Despite his deep respect for established authority, including that of the Church [a point which is arguable, to say the least!], Darwin's theories ran counter to the traditional rationale for such authority. Ffinden did indeed write to [Sir John]

> Lubbock of the 'harmful tendencies to the cause of revealed religion of Mr Darwin's views', adding that he trusted 'that God's grace might in time bring one so highly gifted intellectually & morally to a better mind'.[18] Yet such theoretical differences, though manifest, had not divided Darwin and Innes, who shared Ffinden's High Church training and outlook.[19]

In May 1872 Darwin sent Ffinden a cheque for £35 'as his subscription towards the building of a vicarage'.[20]

> Darwin sent Ffinden 'contributions to the Down Coal and Clothing Club, consisting of £5 for self, £1 for [i.e. on behalf of] my son George, & £3 for my son Francis' in December 1877.[21]

In late November 1873, Darwin thanked Innes for sending him a copy of a sermon by theologian Dr Edward B. Pusey (which, on the 3rd of that month, had been read by theologian Henry P. Liddon at the Church of St Mary's, Oxford, in which Pusey attacked the theory of evolution. Darwin commented:

> I am a little disappointed in it, as I expected more vigour & less verbiage. I hardly see how religion & science can be kept as distinct as he desires, as geology has to treat of the history of the Earth & Biology that of man. But I most wholly agree with you that there is no reason why the disciples of either school should attack each other with bitterness, though each upholding strictly their beliefs.

For his part, Darwin could not 'remember that I have ever published a word directly against religion or the clergy.'[22]

The following day, to botanist Henry N. Ridley, Darwin wrote:

> I have never answered criticisms excepting those made by scientific men … . Dr Pusey's attack will be as powerless to retard by a day the belief in evolution as were the virulent attacks made by divines fifty years ago against Geology, & the still older ones of the Catholic church against Galileo, for the public is wise enough always to follow scientific men when they agree on any subject; & now there is almost complete unanimity amongst Biologists about Evolution … .[23]

Darwin's annual contributions to the church dropped from £50 in 1872 to £10 in 1873, and less thereafter; but he continued to give large sums for restoration work.[24]

Towards the end of 1880 Darwin agreed that 'the young Brethren evangelist' James Fegan should have the use of Downe's aforementioned reading room. Wrote he to Fegan:

> May I have the pleasure of handing the Reading Room over to you? [i.e. to rent] … you have done more for the village in a few months than all our efforts for many years. We have never been able to reclaim a drunkard, but through your services I do not know that there is [one] left in the village.[25]

Members of Darwin's family sometimes attended Fegan's services, altering their dinner hour to do so.[26]

Chapter 41

The Darwin Children

Of Darwin's love for his children there is no doubt. For them to be happy was his earnest wish, and one may imagine them playing tennis or croquet on the lawn, accompanying him on nature study walks, or propelling themselves down the wooden slide which he had installed especially for their amusement on the short staircase leading from the half-landing. However, this did not preclude him from regarding them as appropriate subjects for scientific study. As a father he wore two hats, firstly as a loving and protective parent, sensitive to their sufferings, a fact which made him intensely vulnerable in particular during their periods of illness; secondly as a scientist, who recorded their every gesture, grimace, sign of pleasure or displeasure, how they interacted with him and with other family members, and how their antics resembled those of the apes, to whom, as he proved, they were closely related.

For example, in respect of (William) Erasmus, born 27 December 1839, he describes how

> At his 8th day [i.e. when he was seven days old] he frowned much … now his eyebrows are very little prominent & scarcely a vestige of down [fine soft hair on the face or body[1]]: therefore if frowning has any relation to vision, it must now be quite instinctive: moreover vision at this age is exceedingly imperfect. At his 9th day, however, he appeared to follow a candle with his eyes.[2] When 7-weeks old, his eyes were attracted by a dangling tassle & a bright colour.[3]

When Erasmus was aged four months, commented Darwin,

> I made [a] loud snoring noise near his face, which made him look grave & afraid & then suddenly burst out crying. This is curious, considering the wondrous number of strange noises, & stranger grimaces I have made at him, & which he has always taken as a good joke. I repeated the experiment.[4]

And when Erasmus was aged one year, he 'kissed himself in the glass [mirror] & pressed his face against his image very like [Orang-utan]'.

Darwin made similar observations about his other children – the way they expressed pleasure or pain; their progress in learning to speak; the development of their characters; how they both gave and demanded affection, etc. (Such observations by Darwin culminated in his book *The Expression of the Emotions in Man and Animals*, published in 1872.)

A caring, but not uncritical, father

When daughter Anne was three years old, Darwin wrote, 'Obstinacy is her chief fault at present.'[5]

To his eldest son William, in October 1851, Darwin wrote of his second son, George Howard Darwin, now aged six:

> All day long Georgy is drawing ships or soldiers, more especially drummers whom he will talk about as long as anyone will listen to him.[6]

When, in early February 1852, William entered Rugby School as a boarder, Darwin wrote to him to say, 'We are so very glad to hear that you are happy & comfortable; long may you keep so my dear Boy.'[7] However, the following year he issued this caveat, in regard to schools in general:

> my main objection to them, as places of education, is the enormous proportion of time spent over classics. I fancy, that I can perceive the ill & contracting effect on my eldest Boy's mind, in checking interest in anything in which reasoning & observation comes into play. Mere memory seems to be worked.[8]

That is, the pupils were not encouraged to think for themselves. It was therefore not surprising that Darwin chose a different establishment for the education of his remaining sons, namely Clapham School, where they would be taught mathematics and science by the Reverend Charles Pritchard, a contemporary of his at Cambridge.

The children's health

Of the Darwins' ten children three did not survive beyond the age of ten. Anne Elizabeth, born in 1841, died in 1851 (allegedly) of tuberculosis;[9] Mary Eleanor, born in 1842, died aged twenty-three days of an unknown disease; and Charles Waring, born in 1856, died aged eighteen months of scarlet fever. (According to Randal H. Keynes, Charles may also have been a sufferer from Down's syndrome.)[10]

In his correspondence, Darwin describes the symptoms from which Anne suffered. For example, from Malvern, in March 1851, he wrote to Fox to say,

> I have brought my eldest girl here & intend to leave her for a month under Dr Gully; she inherits, I fear with grief, my wretched digestion.[11]

This suggestion by Darwin that Anne may have inherited the illness from which he himself suffered was presumably based on the fact that his daughter's symptoms were similar to his own, notably in respect of her vomiting. This is discussed below.

The following month, Darwin wrote to Emma from Malvern, in respect of Anne:

> she keeps the same, quite easy, but I grieve to say she has vomited a large quantity of bright green fluid. Her case seems to me an exaggerated one of my Maer illness.[12]

Here Darwin is referring to the fact that he himself had fallen seriously ill at Maer Hall, home of Josiah Wedgwood II, on 4 August 1840 and had not returned to his own home until 14 November.[13]

Anne died at Malvern on 23 April 1851, aged ten years. Six days later, Darwin wrote again to Fox. 'Thank God she suffered hardly at all, & expired as tranquilly as a little angel.'[14]

Darwin agonizes over whether he was in any way to blame for the ill health of his children

Apart from those mentioned above who died young, several of the Darwins' surviving children experienced ill health, notably Henrietta, Leonard, Horace, and Erasmus. This led Darwin to declare, in April 1858,

> I have now six Boys!! & two girls; & it is the great drawback to my happiness, that they are not very robust; some of them seem to have inherited my detestable constitution.[15]

Evidence has been produced that Darwin's chronic illness was the result of Chagas' disease. However, although this is an infectious disease, it is virtually impossible for a person to contract it from another (infected) person, for reasons already discussed. Therefore, Chagas' disease was not the culprit.

(It crossed the mind of the author that Darwin may have brought back some Reduvius 'bugs' from South America, but London's Natural History Museum has no record of this.)

Darwin's investigation of cousin marriages

Darwin was aware not only that there had been intermarriages between Darwins and Wedgwoods, but also between Wedgwoods and Wedgwoods. For example, he himself had married his first cousin Emma, and Emma's grandfather Josiah Wedgwood (I) and Josiah's wife Sarah (née Wedgwood), were third cousins. Therefore, when, on 17 July 1870, he wrote to his near neighbour Sir John Lubbock MP, the possibility that such 'inbreeding' had been the cause of his children's illnesses was clearly in his mind.

> As I hear that the census will be brought before the House to-morrow [a reference to the Second Reading of the Census Bill by the House of Commons], I write to say how much I hope that you will express your opinion on the desirability of queries in relation to consanguineous [relating to or denoting people descended from the same ancestor][16] marriages being inserted. As you are aware, I have made experiments on the subject during several years; and it is my clear conviction that there is now ample evidence of the existence of a great physiological law, rendering an enquiry with reference to mankind of much importance. In England, & many parts of Europe the marriages of cousins are objected to from their supposed injurious consequences; but this belief rests on no direct evidence. It is therefore manifestly desirable that the belief should either be proved false, or should be confirmed, so that in this latter case the marriages of cousins might be discouraged.[17]

In other words, by analyzing data produced in this way by the census, the question would be answered once and for all. Lubbock did attempt to have the Census Bill amended along the lines which Darwin had suggested but, this being a politically sensitive subject, his amendment was defeated. This led Darwin to declare, angrily,

> When the principles of breeding and inheritance are better understood, we shall not hear ignorant members of our legislature rejecting with scorn a plan for ascertaining whether or not consanguineous marriages are injurious to man.[18]

Recent research into the subject

In May 2010 scientists Tim M. Berra, Gonzalo Alvarez, and Francisco C. Ceballos published the results of their study entitled 'Was the Darwin/Wedgwood Dynasty Adversely Affected by Consanguinity?'[19]

A genetic explanation

It has always been known that consanguineous unions may have a deleterious effect on the offspring of such unions. However, only since Darwin's time has an explanation been forthcoming.

In the offspring of consanguineous unions there is a far greater likelihood than would otherwise be the case of alleles occurring at a particular locus of the chromosome being identical. Furthermore, if both of these identical alleles are recessive, then the recessive gene produced in this way will appear in the genotype (genetic constitution) of the offspring, and because recessive genes tend to code for traits which are less advantageous to the individual than is the case for dominant genes, this will tend to have a negative affect on that individual's phenotype – set of observable characteristics or traits.[20]

Having made an exhaustive study of the Darwin/Wedgwood pedigree, the authors concluded that yes, 'the high childhood mortality experienced by the Darwin progeny might be a result of consanguineous marriages within the Darwin/Wedgwood dynasty'.

Could Darwin have infected his children with Chagas' disease?

Person-to-person transmission of Chagas' disease is not thought to be possible, in normal circumstances. However, researchers Luiz Carvalho and others have discovered, from experimentation with mice, that 'T. cruzi is able to reach the seminiferous tubule lumen, thus suggesting that Chagas' disease could potentially be transmitted through sexual intercourse'.[21]

But if it was the case that Darwin's semen did contain T. cruzi and if he transmitted this organism to his wife Emma during sexual intercourse, this does not explain why she herself did not contract the disease, or why seven of the Darwins' ten children were apparently healthy.

The Darwins' surviving children

William Erasmus. Born on 27 December 1839 and educated at Christ's College, Cambridge, he became a banker in Southampton, Hampshire. He married Sara Price Ashburner of New York. There were no offspring. He died in 1914.

Henrietta Emma ('Etty'). Born on 25 September 1843, she married Richard Buckley Litchfield. Henrietta edited her mother Emma's personal letters and had them published in 1904. There were no offspring and she died in 1929.

George Howard. Born on 9 July 1845 and educated at St John's and Trinity College, Cambridge, he became a mathematician, barrister-at-law, Plumian Professor of Astronomy and Experimental Philosophy at Cambridge University and Fellow of the Royal Society. George was knighted in 1905. He married Martha (Maud) du Puy from Philadelphia, and the couple had two sons and two daughters. He died in 1912.

Elizabeth. Born on 8 July 1847, she did not marry. She died in 1926.

Francis. Born on 16 August 1848, he was educated at Trinity College, Cambridge and St George's Hospital Medical School, London, and became Professor of Botany at Cambridge and a Fellow of the Royal Society. Francis was knighted in 1913. He edited much of his father's correspondence, and published the *Life and Letters of Charles Darwin* in 1887 and *More Letters of Charles Darwin* in 1903. He also edited and published Darwin's *Autobiography*. His first marriage was to Amy Ruck, who died shortly after the birth of their son Bernard. His second wife was Ellen Crofts, by whom he had a daughter Frances. He subsequently married Florence H. Fisher. Francis died in 1925.

Leonard. Born on 15 January 1850, he served as a major in the Corps of Royal Engineers, taught at the School of Military Engineering, Chatham, and served in the War Office, Intelligence Division. He became Liberal-Unionist MP for Lichfield, Staffordshire, was a Fellow of the Royal Society, and President of the Royal Geological Society from 1908 to 1911. He was married to Elizabeth F. Fraser, and subsequently to Charlotte M. Massingberd. There were no offspring. Leonard died in 1943.

Horace. Born on 13 May 1851, he was educated at Trinity College, Cambridge and became an engineer and designer of scientific instruments. He was co-founder of the Cambridge Scientific Instrument Company, Mayor of Cambridge, and Fellow of the Royal Society. Horace was knighted in 1918. He married Emma C. Farrer, who bore him three children. He died in 1928.

Chapter 42

The Final Decade

Darwin's *Expression of the Emotions in Men and Animals* was published in 1872.

The following January, Emma wrote to her aunt Fanny Allen to say that she was shortly to meet the Reverend Moncure Daniel Conway. 'We have just been reading a very grand sermon of his on Darwinism.'[1] In fact, Darwin and Conway were acquainted, as will shortly be seen.

Born in Virginia, USA in 1832, Moncure Daniel Conway was the son of an attorney, judge, and slave-plantation owner. He himself was an amateur anthropologist who enjoyed studying the customs and religions of the various races of the world. In 1854, when he became a Unitarian minister, Conway preached anti-slavery sermons and, in 1858, he 'debunked New Testament miracles'.[2] In 1863 he came to England and, the following year, became the preacher at South Place (Unitarian) Chapel and leader of the South Place Religious Society (SPRS), Finsbury, London.

The SPRS had its origins in an earlier society, founded in 1793 by a group of nonconformists, known as 'Philadelphians' (from the Greek *philadelphia* – 'brotherly love') or 'Universalists' (a Universalist being a member of an organized body of Christians who hold the belief that all humankind will eventually be saved[3] – i.e. that salvation is available to all). In June 1819 the SPRS formally united itself with the Unitarian Association.[4]

On 22 May 1823 the SPRS's then minister William Johnson Fox, laid the foundation stone of the South Place Chapel, Finsbury, London, which was duly opened on 1 February the following year. A decade later, the Society severed its connections with Unitarianism, since when it has been independent of any other denomination.

In his address to the Society, delivered on 27 March 1842, Fox declared, 'I Believe in the duty of free enquiry, and in the right of religious liberty.'[5] He then asked:

> What is Deity? Let creeds tell us what they please, in phraseology simple or obscure – what is Deity but the loftiest conception of each mind? As high as each soul can get in its notion of the true, the wise, the good, the powerful – that, to each, is God. All else is verbiage.
>
> The duty of free enquiry, or 'the right of religious liberty,' – what means it

> but that without let, molestation, or hindrance, without censure, aversion, or punishment, man is to think whereto his being tends?[6]

Despite his assertion that God was man made, rather than the other way around, Fox declared that it was important 'to derive all the truth and beauty that we can from the scriptures of the Old and the New Testaments'.[7]

Therefore, Fox's notion that God is a product of the human mind coincided with Darwin's own beliefs. Finally, in 1888, the SPRS was renamed the South Place Ethical Society.[8]

In his *Centenary History of the South Place Society*, published in 1894, Conway refers to a

> Discussion Society, formed in 1877 … [and to] the 'Conference of Liberal Thinkers', which gathered here in 1878; and the many course of lectures, beginning with those of 1873, when … Huxley, and the younger Darwin [i.e. Charles] were heard here [i.e. addressed the Society], and extending to those on 'Religious Systems of the World' and 'National Life and Thought'.[9]
>
> I recall with happiness that Charles Darwin expressed to me his warm interest in South Place, as did Sir Charles and Lady Lyell, whom we often saw here; and [philosopher and social reformer] John Stuart Mill and Professor [William Kingdom] Clifford [Professor of Applied Mathematics at University College, London].[10]

Conway demonstrated that he possessed a more than rudimentary grasp of the meaning of Darwinism, when he declared that:

> All moral or social progress depends on the freedom to develop differences. In every step of physical progress, from the sponge up to man, we see a small difference of some organism from its species, and a long struggle of that difference to hold its own against the majority, and gradually develop a new species with its variation from the old.

Conway, whose inclination was now towards humanism and freethinking, extended this analogy to religion. It was

> by differentiation of religious belief [that] larger religions were evolved. And so on with all the various developments, whose sum we call Civilization.
>
> Now, one fundamental fact we can see which our freethinking fathers could not see, namely, that all moral or social progress depends on the freedom to develop differences. That is what the great discovery of Darwin taught us.

> Freedom to differ from the majority is, therefore, the condition [i.e. prerequisite] of all progress, social, moral, and physical.[11]

Darwin's lifelong preoccupation had been with how animals and plants evolve physically, but he would undoubtedly have agreed with Conway that the human mind is also capable of evolving, and will do so more readily in a free environment.

Conway continued, scathingly:

> How many great works of genius have been burnt by the common hangman? Nearly every one up to the Reformation, and many after that. The authority of a few particularly unfit officials to decide for forty million English intellects what they shall read, appears intolerable … .[12]
>
> Antiquated laws are sometimes spoken of as 'dead letters' but they are never dead; they continue their subtle work in the mind of the people, and survive as the prejudices which discourage the thinker and retard progress. It is, therefore, a great and worthy aim to which our energies may be wisely consecrated, in this centenary year of our Society, to obtain the removal of all surviving censorships on literature, art, and ethical science, maintain the honour of individual independence, and establish entire intellectual, moral, and religious Liberty.[13]

Again, Darwin would have concurred with such sentiments absolutely.

To Nicolaas D. Doedes, student of natural history at the University of Utrecht, Holland, Darwin declared, in April 1873,

> the impossibility of conceiving that this grand and wondrous universe, with our conscious selves, arose through chance, seems to me the chief argument for the existence of God; but whether this is an argument of real value, I have never been able to decide. The safest conclusion seems to be that the whole subject is beyond the scope of man's intellect … .[14]

The following September Darwin wrote frankly to Lyell on the same subject.

> Many persons seem to make themselves quite easy about immortality & the existence of a personal God by intuition; & I suppose that I must differ from such persons, for I do not feel any innate conviction on any such points.[15]

Returning to the theme in March 1878, to Scottish journalist James Grant, Darwin opined that:

> The strongest argument for the existence of God, as it seems to me, is the instinct or intuition which we all (as I suppose) feel that there must have been

> an intelligent beginner of the Universe; but then comes the doubt and difficulty whether such intuitions are trustworthy.[16]

And that November, he told botanist Henry N. Ridley:

> many years ago when I was collecting facts for the *Origin*, my belief in what is called a personal God was as firm as that of Dr Pusey himself, & as to the eternity of matter I have never troubled myself about such insoluble questions.[17]

In April 1879 Emma wrote to Russian diplomat Baron Nicolai A. von Mengden to say of her husband:

> He considers that the theory of evolution is quite compatible with the belief in a God; but that you must remember that different persons have different definitions of what they mean by God.[18]

That is to say, Emma was not convinced that the God whom Darwin had in mind was the same one whom she as a Christian worshipped.

In May, to John Fordyce, author of works on the modern social order, Darwin wrote:

> It seems to me absurd to doubt that a man may be an ardent Theist & an evolutionist. What my own views may be is a question of no consequence to any one except myself.— But as you ask, I may state that my judgement often fluctuates. Moreover whether a man deserves to be called a theist depends on the definition of the term: which is much too large a subject for a note. In my most extreme fluctuations I have never been an atheist in the sense of denying the existence of a God. I think that generally (& more and more so as I grow older) but not always, that an agnostic would be the most correct description of my state of mind.[19]

(Atheist – a person who does not believe in the existence of God or gods. Agnostic – a person who believes that nothing is known or can be known of the existence or nature of God or of anything beyond material phenomena; a person who claims neither faith nor disbelief in God.)[20]

> To Frederick A. McDermott, barrister, Darwin wrote in November 1880:
> I am sorry to have to inform you that I do not believe in the Bible as a divine revelation, & therefore not in Jesus Christ as the son of God.[21]

The only possible conclusion to be reached from this statement by Darwin, is that he finds the notion of the *Holy Bible* as a source of revelation – defined as a divine or supernatural disclosure to humans of something relating to human existence or the world[22] – completely unacceptable. Therefore, despite admitting that his 'judgement often fluctuates', and despite declaring himself to be an agnostic, he was indisputably, at any rate as far as the Christian religion is concerned, an atheist, although this did not preclude him from believing in 'a God' (as already mentioned) – though he was not specific about whosoever this deity might be.

For Darwin retirement was not an option and he continued with his scientific researches almost to the very end, as this letter, printed in the correspondence column of *The Academy* (a weekly review of literature and general topics) reveals. Its author was one William Watson of Ormskirk, Lancashire and the subject, surprise, surprise, was natural history!

> It has been suggested to me that a letter which I received from Mr Darwin on the day before he died, though not important in itself derives from the accident of being among the latest things he wrote, an interest such as entitles it to publicity. Written by return of post in answer to the mere casual communication of a stranger, it has, at all events, the interest of being one of the many illustrations of that almost proverbial courtesy which characterized the greatest, since Newton, of 'those who know'. I had taken the liberty of pointing out to him what seemed to me, for certain reasons, a false conclusion arrived at in a paragraph of 'The Expression of the Emotions in Man and Animals', where Darwin certainly seems to imply that the familiar canine practice of throwing up earth by backward ejaculations of the hind-feet is a 'purposeless remnant' of a habit, on the part of the dog's wilder progenitors, of 'burying superfluous food'. Mr Darwin's reply was as follows:
>
>> Dear Sir,
>> You have misunderstood my meaning; but the mistake was a very natural one, and your criticism good. I ought not to have interpolated the sentence about the burying of food; and if inserted at all, it ought to have been at end of [the] paragraph, or in a separate one. The case was instanced solely to illustrate a long-continued habit, for, as far as I have seen, well-fed domestic dogs do not revisit their buried treasures. A dog when burying food makes a hole (as far as I have seen) with his front legs alone, and thrusts in the earth with his nose, so that there is no resemblance to the supposed excrement-covering movements,
>> Dear Sir, Yours faithfully, CH. DARWIN.[23]

Darwin died at Down House on 19 April 1882 at the age of seventy-three. It had been his wish that his funeral should take place at the Church of St Mary the Virgin, Downe, and that he should be buried in the churchyard, alongside his children who had predeceased him and his brother Erasmus, who had died the year before. However, Sir John Lubbock 'drew up the memorial [a statement of facts, especially as the basis of a petition][24] to the Dean of Westminster', requesting that Darwin be buried in Westminster Abbey.[25]

Said his friend and colleague A. R. Wallace, 'I was honoured by an invitation to his funeral in Westminster Abbey, as one of the pall-bearers ….'[26] Other pall bearers included Huxley and Lubbock.

The funeral sermon was conducted by the Reverend Frederic W. Farrer, Master of Harrow School, whom Darwin had once proposed for fellowship of the Royal Society in recognition of his work on philology. For Darwin, the irony of his being buried in Westminster Abbey, the mighty monument to Anglicanism, would not have been lost on him!

A memorial service was subsequently held at South Place (Unitarian) Chapel, Finsbury, London in Darwin's honour, and conducted by the Reverend Moncure Conway.[27] Conway was later to say, proudly, that among the 'veteran Freethinkers and Radicals at whose graves I officiated were Sir Charles Lyell [who had died in 1875] and Charles Darwin'.[28]

Said the Darwins' daughter Henrietta Litchfield:

> My mother spent the summer of 1882 at Down, but she felt that the winters in the great empty house would be lonely, and she therefore decided to spend part of each year at Cambridge, where two of her sons, George and Horace, were living, and where her son Francis could better go on with his botanical work.
>
> She therefore bought 'The Grove', a pleasant house on the Huntingdon Road, a mile from great St Mary's, and there she spent the winters till her death.[29]

Henrietta also wrote:

> The Memorial Statue of my father was unveiled on the 9th June, 1885 at the Natural History Museum. My mother did not attend the ceremony; she wrote, 'I should like very much to be present but I should prefer avoiding all greetings and acquaintances.'[30]

Emma died on 2 October 1896. She is buried at Down.

Epilogue

Darwin, the reticent atheist

In the early years of his disillusionment with religion, and Anglicanism in particular, Darwin was relatively reticent about what he described, euphemistically, as his 'agnosticism' – which was, in reality, atheism. This behaviour on his part was undoubtedly primarily in order to protect the feelings of his wife Emma. However, as time went by, he became more and more outspoken.

Dawkins and the invidious nature of religious indoctrination

As already mentioned, Darwin had grave reservations about religious indoctrination, especially of children. Professor Dawkins is of the same opinion.

> I don't think it's harmless. There is something insidious about training children to believe things for which there is no evidence.[1]

Had Darwin been alive today …

Darwin, in his bachelor days, had composed a list of the pros and cons of marriage. What if he had composed a similar list of the pros and cons of living in the twenty-first century? Perhaps it would have read something like this.

a. Sources of satisfaction, wonder and delight

Darwin would have marvelled at how:

i. Mankind has benefited from modern medicines and vaccines.
ii. Modern agricultural methods are helping to feed the world.
iii. Space exploration is now a reality and, with it, the prospect of discovering how the universe has evolved.
iv. Fossils of previously unknown species continue to be discovered and are taking their place on the evolutionary tree of life.
v. It has recently been demonstrated that, as he and Lamarck always suspected, the environment *does* have an influence on the way variations of species come about, and the mechanism by which this occurs has now been demonstrated, though this branch of science is as yet in its infancy. Furthermore, it has been shown that the environment can have an influence, not only on the morphology

of an organism, but also, and in particular in the case of man, on his behaviour, mood, belief systems, etc., and also on those of his descendants.
vi. The truth of his great theory of evolution has been universally accepted by the scientific community, and fully corroborated by the science of genetics.

b. Sources of disquiet
On the other hand, he would have been alarmed that:

i. The Reverend Thomas Malthus's dire warnings about the perils of overpopulation, delivered more than two centuries ago, are still being ignored; meanwhile, the population of the world continues to increase virtually unchecked. Today, Malthus's views are echoed by naturalist and broadcaster Sir David F. Attenborough, who declared:

> The population of the world is now growing by nearly 80 million a year. One and a half million a week. A quarter of a million a day. Ten thousand an hour. In this country it is projected to grow by 10 million in the next 22 years. That is equivalent to ten more Birminghams.
>
> All these people, in this country and worldwide, rich or poor, need and deserve food, water, energy and space. Will they be able to get it? I don't know. I hope so. But the government's chief scientist and the last president of the Royal Society have both referred to the approaching 'perfect storm' of population growth, climate change and peak oil production, leading inexorably to more and more insecurity in the supply of food, water and energy.
>
> Why this strange silence? I meet no one who privately disagrees that population growth is a problem. No one – except flat-Earthers – can deny that the planet is finite. We can all see it – in that beautiful picture of our Earth taken by the Apollo mission. So why does hardly anyone say so publicly? There seems to be some bizarre taboo around the subject.
>
> It remains an obvious and brutal fact that on a finite planet human population will quite definitely stop at some point. And that can only happen in one of two ways. It can happen sooner, by fewer human births – in a word, by contraception. That is the humane way, the powerful option that allows all of us to deal with the problem, if we collectively choose to do so. The alternative is an increased death rate – the way that all other creatures must suffer, through famine or disease or predation. That, translated into human terms, means famine or disease or war – over oil or water or food or minerals or grazing rights or just living space.
>
> The sooner we stabilize our numbers, the sooner we stop running up the 'down' escalator. Stop population increase – stop the escalator – and we have some chance of reaching the top; that is to say, a decent life for all.[2]

ii. Human activity is causing the extinction of species and a consequent reduction in biodiversity.

iii. Again (allegedly) as a result of human activity, global warming is now a serious threat to the planet and all who reside thereon.

iv. Some have even spoken of another 'Great Extinction', in which man, this time, will be the principal victim.

v. There still exist people who, for reasons already discussed, choose to deny the truth of Darwin's great theory.

Professor Steve Jones and *Origin*
In the view of geneticist Steve Jones, Darwin's Theory of Evolution by Natural Selection provides the basic grammar without which it is impossible to understand the science of biology.

> You can't learn a language without understanding at least something about its grammar, and you couldn't be a biologist before 1859 (when *Origin* was first published) because none of the facts seemed to fit together. But suddenly *The Origin of Species* made it all make sense.

Darwin's magnum opus, said Jones, 'really was and still is the central book of the science of biology'.[3]

Professor Dawkins and the solace of Darwinism
Said Professor Dawkins

> There is a sort of happiness, there is a sort of bliss, in understanding the elegance with which the world is put together, and Darwinian natural selection is a supremely elegant idea. It really does make everything fall into place and make sense, and I find great consolation, great happiness, in that level of understanding.[4]

Some may indeed find these concepts consoling. For others, however, this impersonal explanation of life will not be enough. For them, it is impossible to accept that there is no ultimate purpose in the existence of human beings, or of other animals, plants, the Earth, or the Universe, and they will therefore cling on, come what may, to their gods, whether benign and seemingly reasonable, or bloodthirsty, intolerant, or irrational, and they will do so however uncorroborated the evidence for the existence of these gods may be. Another powerful influence is the comfort factor. How much more consoling it is to believe that there is a god who cares about us, especially in adverse circumstances.

Down House

Here we may imagine Darwin in his prime, pottering about in the hot house amongst his precious plant specimens; rummaging in the undergrowth collecting beetles to swell his collection, or light-heartedly pushing his children on their garden swing, as his beloved terrier snaps playfully at his heels.

Darwin's former home remained in the Darwin family until 1927. It subsequently became a girls' school. English Heritage is now its proprietor and it is open to the public.

Darwin resides in splendour

On 23 May 2008 it was announced that the 2.2 tonne statue of Charles Darwin was to be replaced in its original position – at the top of a flight of stairs in the Central Hall of London's Natural History Museum.

> The Darwin statue was created by Sir Joseph Boehm and was unveiled on 9 June 1885. In 1927 it was moved to make way for an Indian elephant specimen [how it would have amused Darwin, to learn this fact!], and then moved again in 1970 to the North Hall.

This was in preparation for 'Darwin200', a nationwide programme of events in 2008/9, celebrating Darwin's ideas, impact and influence, and scheduled to take place at about the time of the bicentenary of his birth.

> Moving the Darwin statue took 8 people about 26 hours. The team had to first move a 1-tonne statue of [Professor] Richard Owen, the Museum's founder, to its new position up on the balcony.[5]

Darwin's upstaging of his former opponent Owen in this way, would undoubtedly have afforded the former a quiet satisfaction whilst enraging the latter! So now Darwin sits in pride of place in Britain's great monument to science and learning.

What became of HMS *Beagle*?

In 1837, under the command of Commander John C. Wickham, she set off to survey the coast of Australia. Two years later Wickham named a harbour in northern Australia, 'Port Darwin', in honour of *Beagle*'s former shipmate. Here, a settlement developed into what became the town of Palmerston, which was likewise renamed Darwin, in 1911.

EPILOGUE

In 1845 *Beagle* was used by HM Customs and Excise as a watch vessel, and moored in the river Roach to control smuggling on the coast of Essex. In 1870, at the end of her working life, she was allegedly broken up, or possibly sunk deliberately in the Roach. How sad that it was not possible to restore her to her former glory and put her on display as a national treasure.

Darwin's legacy

In addition to demonstrating how all life on Earth, mankind included, and both the living and the extinct, has evolved, Darwin's legacy was to encourage each and every one of us to think for ourselves: a process which is not only extremely liberating, but which can also be a source of enormous joy and fulfilment, as he himself so amply demonstrated.

Appendix I

HMS *Beagle*'s ship's company, at the time of embarkation for South of her embarkation for South America, 27 December 1831. Included amongst the of 1831. Included amongst the seventy-four persons aboard were the following:

Robert FitzRoy	Commander and surveyor
John Clements Wickham	Lieutenant
Bartholomew James Sulivan	Lieutenant
Edward Main Chaffers	Master
Robert Mac-Cormick	Surgeon
George Rowlett	Purser
Alexander Derbishire	Mate
Peter Benson Stewart	Mate
John Lort Stokes	Mate and Assistant Surveyor
Benjamin Bynoe	Assistant Surgeon
Arthur Mellersh	Midshipman
Philip Gidley King	Midshipman
Alexander Burns Usborne	Master's Assistant
Charles Musters	Volunteer 1st Class
Jonathan May	Carpenter
Edward H. Hellyer	Clerk.

Acting boatswain: Sergeant of marines and seven privates: Thirty-four seamen and six boys

On the List of supernumeraries were –

Charles Darwin	Naturalist.
Augustus Earle	Draughtsman.
George James Stebbing	Instrument Maker.

Richard Matthews and three Fuegians: My own steward: and Mr Darwin's servant.[1]

Appendix II

A selected list of people with whom Darwin corresponded over the years.

1855
Edward Blyth, Curator of the Museum of the Asiatic Society of Bengal.

1858
George Robert Waterhouse, keeper, Department of Geology, British Museum; Frederick Smith, entomologist in the Zoological Department of the British Museum; Charles Cardale Babington, botanist and archaeologist of St John's College, Cambridge; Henry Holland, physician and kinsman of the Darwins.

1859
Andrew Crombie Ramsay, lecturer at the government School of Mines.

1860
Maxwell Tylden Masters, physician and lecturer at St George's Hospital; Thomas Bridges, a missionary at Keppel Island in the West Falklands; Hugh Falconer, palaeontologist and botanist; Samuel Pickworth Woodward, First Class Assistant in the Department of Geology and Mineralogy, British Museum; William Bernhard Tegetmeier, editor, journalist, lecturer, naturalist, pigeon fancier, and an expert on fowls; William Henry Harvey, botanist and keeper of the Herbarium, Trinity College, Dublin; William Hallowes Miller, Professor of Mineralogy at Cambridge.

1861
Henry Walter Bates, naturalist, who had undertaken a joint expedition to the Amazon with Alfred R. Wallace between 1848 and 1850; Dorothy Fanny Nevill of Sussex, cultivator of tropical plants; Jeffries Wyman, US anatomist and ethnologist.

1862
Edward Cresy, surveyor and civil engineer; George Bentham, botanist and President of the Linnaean Society; Charles Kingsley, author, clergyman, and Professor of Modern History at Cambridge University; Alexander Goodman More, naturalist; Thomas G. Appleton, Boston, essayist, poet and artist; Thomas Rivers, nurseryman and collector of roses.

1863
Thomas Rivers, nurseryman and a founder of the British Pomological Society (pomology being the science of fruit growing); John Scott, Scottish botanist and foreman of the Propagating Department at the Royal Botanic Garden, Edinburgh; John Evans, paper manufacturer, archaeologist, geologist, and numismatist (student of coins and medals); Darwin's son, George Howard Darwin; Francis T. Buckland, naturalist, popular science writer and surgeon.

1864
Daniel Oliver, keeper of the herbarium at the Royal Botanic Gardens, Kew and Professor of Botany at University College, London; William Erasmus Darwin, Darwin's eldest son, who was a partner in the Southampton and Hampshire Bank, chairman of the Southampton Water Company, amateur photographer, and amateur botanist.

1865
Thomas Campbell Eyton, Shropshire naturalist; William Bowman, a leading ophthalmic surgeon; William Charles Linnaeus Martin, former Superintendent of the Zoological Society of London and writer on natural history.

1866
Philip Lutley Sclater, lawyer and ornithologist; Johann F. T. Müller, German naturalist; Swiss botanist Carl Wilhelm von Nägeli.

1867
Johann F. T. Müller, German naturalist; Thomas Belt, geologist, naturalist and mining engineer; Julius V. Carus, German comparative anatomist; James P. M. Weale, naturalist, farmer, and writer; William Bowman, ophthalmic surgeon; (Felix) Anton Dohrn, German zoologist; John Scott, Scottish botanist.

1868
John Jenner Weir, naturalist and accountant; Roland Trimen, zoologist and entomologist; John Scott, Scottish botanist; Friedrich Rolle, German geologist and palaeontologist.

1869
Dr Henry Maudsley, doctor, psychologist and psychiatrist; James Crichton–Browne, Scottish physician and psychiatrist; William Ogle, physician and naturalist; Frederico Delpino, Italian botanist; Albrecht C. L. G. Günther, German-born zoologist.

APPENDIX II

1870

Vladimir O. Kovalevsky, Russian palaeontologist; Patrick Nicol, Scottish-born physician and mental-health practitioner; Bartholomew J. Sulivan, Darwin's officer/hydrographer and companion on HMS *Beagle*; William Swale, nurseryman who emigrated to New Zealand; John Lubbock, 1st Baron Avebury, banker, politician, and naturalist.

1871

St George Jackson Mivart, comparative anatomist; W. J. Erasmus Wilson, dermatologist and philanthropist; Joseph Wolf, German-born painter and illustrator; Alexander Agassiz, Swiss-born geologist, oceanographer, and mining engineer; Thomas Henry Farrer, 1st Baron Farrer, civil servant and barrister; George Busk, Russian-born naval surgeon and naturalist; Osbert Salvin, ornithologist and entomologist.

Appendix III

A selected list of topics which interested Darwin over the years.

1858
Variations in birds' nests; 'humble-bees', and the bees of New Zealand, and how 'Esquimaux who have not iron [tools] cut holes in ice for catching fish & Seals, & dig snow-houses &c.'

1860
'Whether there is much variation in Sweet Peas which might be owing to natural crosses'; whether it was 'possible that archaeologists may know when our gigantic dray or wagon Horses were first recorded or noticed'; those places 'where the Bee or Fly Orchis is tolerably common …'; the '…exact dates of period [length] of gestation of any breeds of Dog?'.

1861
How the skulls of various types of fowl differed; 'whether the glacial period affected the whole world contemporaneously …'; how variations in hollyhocks had come about; the means by which orchids are pollinated, and how the ancient Scottish lakes of Lochaber had been formed.

1862
The fertilization of orchids; whether insects change quickly with time; melastomas (tropical shrubs and small trees from South-East Asia), and Drosera (carnivorous plants, otherwise known as sundews).

1863
The cultivation of peaches; the ability of a fish to re-grow its fin after it had been severed; 'the proportion of men who suffer from Tropical diseases in relation to the colour of their hair & skin', and the cross-breeding of primroses and cowslips.

1865
The webbed feet of otter hounds, as compared with fox hounds or harriers; the speed at which pigeons fly, and variations to be found in the plumage of birds.

APPENDIX III

1866
The incidence of squinting (strabismus) in children; the architecture of the honeycombs of bees; the colouration of the wings of butterflies, and the cross-breeding of cucumbers.

1867
Darwin's current preoccupation was with botany, but other subjects of interest included facial expressions exhibited by various races; why caterpillars are 'sometimes so beautifully and artistically coloured?'; horned beetles, and why the female birds of certain species are more brightly coloured than the males.

1868
Possible affinities between the languages of Australia and those of Africa; the proportion of males to females to be found in the various species of the animal kingdom; whether butterflies are polygamous; how it can be that 'a single cell of a plant or the stump of an amputated limb have the "potentiality" of reproducing the whole …', and the factors which influence the production of tears, not only in humans but also in elephants!

1869
The origin of the inhabitants of New Zealand; 'the plumage of the chickens of Cuckoo breeds & Seabright bantams'; the 'Expression of the Emotions amongst the insane and idiotic', and which of the mammalian species are polygamous.

1870
Suicide amongst 'savages'; body language in respect of savages and Cistercian monks; the deaf and dumb; cross-pollination of plants, including those from Africa and Europe; the evolution of language, and finally the number of cobwebs to be found in the hedges at Down House!

1871
'How far the disintegration of rocks goes on beneath a continuous bed of turf'; 'The exact [weight] of earth annually cast up by earth worms over a square yard'; 'The pouting of children of savages, & … the pouting of English children', and the ears of hedgehogs compared with those of man.

Notes

CHAPTER 1

1. Mansion House, called 'The Mount', Shrewsbury: Particulars of Sale By Auction, 1867. From Boyd, P. D. A., 1999, Darwin Garden Project, Shropshire Parks and Gardens Trust Newsletter.
2.Darwin, Francis, *The Life and Letters of Charles Darwin, Including an Autobiographical Chapter*, Volume I, p.19.
3. Darwin, Charles, *The Life of Erasmus Darwin*, p.85.
4. Darwin, Francis, *The Life and Letters of Charles Darwin, Including an Autobiographical Chapter*, Volume I, p.19.
5. Ibid, p.20.
6. Darwin, Francis, *Autobiography of Charles Darwin*, p.7.
7. Ibid, p.2, note 1.
8. Ibid, p.12.
9. Ibid, pp.2-3.
10. Ibid, p.3.
11. Ibid, p.3 and note 2.
12. Ibid, p.5.

CHAPTER 2

1. Oxford Dictionaries Online.
2. Ibid.
3. Reed, the Reverend Clifford M., *Unitarian? What's that?* Para 12.
4. Reed, the Reverend Clifford M., *Beatrix Potter's Unitarian Context*, p.5.
5. Ibid, p.9.
6. Reed, the Reverend Clifford M., *Unitarian? What's that?* Para 38.
7. Martineau, James, *A Seat of Authority in Religion*, 1890, in *James Martineau: Selections*, p.47.
8. Darwin, Francis, *Autobiography of Charles Darwin*, p.2, note 1.
9. Darwin, Charles, *The Life of Erasmus Darwin*, p.63.

CHAPTER 3

1. Darwin, Francis, *Autobiography of Charles Darwin*, p.6.
2. Oldham, J Basil, *A History of Shrewsbury School 1552–1952*, p.99.
3. Oxford Dictionaries Online.
4. Butler, the Reverend Samuel: *Instillation Sermon* at Cambridge, 1811, in J. Basil Oldham, op. cit., p.99.
5. Oxford Dictionaries Online.
6. Darwin, Francis, *Autobiography of Charles Darwin*, p.7.
7. Ibid, p.8.
8. Oxford Dictionaries Online.
9. Darwin, Francis, *Autobiography of Charles Darwin*, p.8.
10. Ibid, p.9.
11. Ibid, p.10.
12. Ibid, p.9
13. Ibid, pp.8–9.
14. Ibid, p.10.
15. Oxford Dictionaries Online.

NOTES

16. E. A. Darwin to Darwin, 14 November 1822, *The Correspondence of Charles Darwin*, Volume 1, 1821–1836, pp.3–4 and note 1, Cor.1, pp.3–4.
17. Darwin, Francis, *Autobiography of Charles Darwin*, p.10.
18. E. A. Darwin to Darwin, 18 May 1823, Cor.1, p.8.
19. E. C. Darwin to Darwin [c. June 1823], Cor.1, p.8.
20. E. A. Darwin to Darwin, 8 March 1825, Cor.1, p.19.
21. Darwin, Francis, op. cit., p.10

CHAPTER 4

1. Darwin, Francis, *Autobiography of Charles Darwin*, p.7.
2. Oxford Dictionaries Online.
3. Darwin, Francis, op cit., p.14.
4. Ibid, p.15.
5. Ibid, p.17.
6. Ibid, p.19.
7. Catherine Darwin to Darwin, 4 December 1825, Cor.1, p.22.
8. Darwin to Caroline Darwin, 6 January 1826, Cor.1, p.25.
9. Darwin to Susan Darwin, 29 January 1926, Cor.1, p.28.
10.Oxford Dictionaries Online.
11. Caroline Darwin to Darwin, 22 March 1826, Cor.1, p.36.
12. Susan Darwin to Darwin, 27 March 1826, Cor.1, p.37.
13. Darwin to W. T. Preyer, 17 February [1870], *The Correspondence of Charles Darwin*, Volume 18, 1870, pp.41–2.
14. Darwin to Caroline Darwin, 8 April 1826, Cor.1, p.39.
15.Caroline Darwin to Darwin, 11 April 1826, Cor.1, p.41.
16. Darwin, Francis, op cit., pp.19–20.
17. Ibid, p.21.
18. Ibid, p.21.

CHAPTER 5

1. Darwin, Francis, op cit., pp.21–3.
2. Ibid, pp.26–7.
3. Ibid, p.25.
4. Darwin to W. T. Preyer, 17 February [1870], Cor.18, p.41.
5. Darwin, Francis, op cit., p.24.
6. Ibid, p.23.
7. Darwin to William Darwin Fox, 12 June 1828, Cor.1, p.56.
8. Darwin to William Darwin Fox, October 1828, Cor.1, p.66.
9. Darwin to Fanny Owen, 26 October 1828, Cor.1, p.69.
10. Darwin to W. D. Fox, 29 October 1828, Cor.1, p.70.
11. 'Mr Dash' was one of at least thirteen dogs which Darwin is known to have possessed, at one time or another, during his lifetime.
12. Darwin to W. D. Fox, 24 December 1828, Cor.1, p.71.
13. Darwin to W. D. Fox, 26 February 1829, Cor.1, pp.75-6.
14. Darwin to W. D. Fox, 15 March 1829, Cor.1, pp.79-80.
15. Darwin to W. D. Fox, 23 April 1829, Cor.1, p.84.
16. Darwin to W.D. Fox, 3 July 1829, Cor.1, p.88.
17. Darwin to W.D. Fox, 9 May 1830, Cor.1, p.103.
18. Darwin, Francis, op cit., pp.26–7.
19. Ibid, p.30.
20. Darwin to Fox, 5 November 1830, Cor.1, pp.109–10.

21. Darwin, Francis, op cit., p.21.
22. Christ's College, Cambridge, Record Book.
23. Darwin to W. D. Fox, 23 January 1831, Cor.1, pp.111–12.
24. Christ's College, Cambridge, Record Book.
25. Oxford Dictionaries Online.
26. Darwin to J. S. Henslow, 11 July 1831. Cor.1, p.125.
27. Darwin, Francis, op cit., p.31.
28. Ibid, p.32.
29. Cambridge University Calendar, 1831, in Cor.1, p.112, and Darwin, Francis, op cit., pp.21–2.

CHAPTER 6

1. Locke, John, *An Essay Concerning Human Understanding*, Book IV, Chapter X.
2. Paley, William, *A View of the Evidences of Christianity*, p.vi.
3. Ibid, p.1.
4. Ibid, p.2.
5. Ibid, p.81.
6. Ibid, p.101.
7. Ibid, p.399.
8. Ibid, pp.403–4.
9. Ibid, pp.134–5.
10. Ibid, p.400.
11. Ibid, p.249.
12. Ibid, pp.264–5.
13. Ibid, p.257.
14. Ibid, p.258.
15. Oxford Dictionaries Online.

CHAPTER 7

1. J. S. Henslow to Darwin, 24 August 1831, Cor.1, pp.128-9.
2. George Peacock to Darwin, C. 26 August 1831, Cor.1, p.130.
3. R. W. Darwin to Josiah Wedgwood II, 30 August 1831, Cor.1, p.132.
4. Darwin to R. W. Darwin, 31 August 1831, Cor.1, pp.132-3.
5. Josiah Wedgwood II to R.W. Darwin, 31 August 1831, Cor.1, pp.133-4.
6. Darwin to Francis Beaufort, 1 September 1831, Cor.1, p.135.
7. Darwin to Susan Darwin [5 September 1831], Cor.1, p.140.
8. Darwin to Susan Darwin, 5 September 1831, Cor.1, pp.140-1.
9. Darwin to Susan Darwin, 6 September 1831, Cor.1, p.143.
10. Darwin to Susan Darwin, 9 September 1831, Cor.1, p.147.
11. Darwin to Susan, 14 September 1831, Cor.1, p.155.
12. Darwin to Caroline Darwin, 12 November 1831, Cor.1, p.178.

CHAPTER 8

1. Darwin, Charles, *The Voyage of the Beagle*, p.15.
2. Cor.1, p.553.
3. Darwin to W. T. Preyer, 17 February [1870], Cor.18, pp.41-2.
4. Cor.1, p.130.
5. Darwin, Charles, op cit., pp.15–16, 21.
6. Darwin to R. W. Darwin, 10 February 1832, Cor.1, p.206.
7. Darwin to R. W. Darwin, 8 February 1832, Cor.1, p.201.

NOTES

8. FitzRoy, Captain Robert, *A Narrative of the Voyage of* H.M.S. *Beagle*, p.50.
9. Darwin, Francis, op cit., p.35.
10. Darwin, Charles, op cit., p.28.
11. FitzRoy, op cit., p.77.
12. Ibid, p.77.
13. Ibid, p.77.
14. Darwin, Charles, *Beagle Diary*, p.173.
15. Darwin to W. D. Fox, May 1832, Cor.1, p.232.
16. Cor.1, p.171.
17. Darwin to J. S. Henslow, 18 May – 16 June 1832, Cor.1, pp. 236–8.
18. Darwin to Catherine Darwin, 5 July 1832, Cor.1, pp.246–7.
19. Darwin, Charles, op cit., pp.64–5, 81.
20. Darwin to J. S. Henslow, 23 July–15 August 1832, Cor.1, p.252.
21. Darwin to J. S.Henslow, 26 October–24 November 1832, Cor.1, p.281.
22. Darwin to Caroline Darwin, 24 October–24 November 1832, Cor.1, pp.276–8.
23. Darwin, Charles, op cit., p.281.
24. Ibid, pp.290–1.
25. Henslow to Darwin, 15–21 January 1833, Cor.1, p.293.
26. Darwin, Charles, op cit., p.260.
27. Darwin to Caroline Darwin 15 March–12 April 1833, Cor.1, p.303.
28. Darwin to J. S. Henslow, 11 April 1833, Cor.1, p.307.
29. Darwin to Catherine Darwin, 22 May–14 July 1833. Cor.1, p.311.
30. Darwin to J. S. Henslow, 18 July 1833, Cor.1, p.322.
31. Darwin, Charles, *The Voyage of the Beagle*, pp.98-9.
32. Ibid, p.151.
33. Ibid, p.100.
34. Ibid, p.112.
35. Ibid, p.119.
36. Ibid, p.119.
37. Ibid, p.120.
38. Ibid, p.211.
39. Ibid, p.145.
40. Ibid, p.147.
41. Ibid, p.169.
42. Ibid, p.173.
43. Ibid, p.184.
44. Ibid, p.187.
45. Ibid, p.220.
46. Ibid, p.220.
47. Ibid, p.230.
48. Ibid, p.235.
49. Ibid, p.236.
50. Oxford Dictionaries Online.
51. Darwin, Charles, op cit., pp.240–2.
52. Ibid, p.321.
53. Ibid, p.324.
54. Ibid, p.325.
55. Ibid, p.310.
56. J. M. Herbert to Darwin 28 March 1834, Cor.1, p.375.
57. Darwin to Catherine Darwin, 6 April 1834, Cor.1, pp.379-80.
58. Darwin, Charles, op cit., pp.245, 248.
59. FitzRoy, Captain Robert, op cit., p.190.
60. Ibid, p.650.

61. Darwin, Charles, op cit., p.330.
62. Ibid, p.335.
63. Darwin to Catherine Darwin, 20 June–9 July 1834, Cor.1, p.391.
64. Darwin, Charles, op cit., p.345.
65. Darwin to Charles Whitley, 23 July 1834, Cor.1, p.397.
66. Darwin, Charles, op cit., p.347.
67. Darwin to Caroline Darwin, 9–12 August 1834, Cor.1, pp.404-05.
68. Darwin to Robert FitzRoy, 28 August 1834, Cor.1, p.406.
69. Darwin, Charles, op cit., p.367.
70. Ibid, p.367.
71. Darwin, Francis, op cit., p.34.
72. Ibid, p.35.
73. Darwin to Catherine Darwin, 8 November 1834, Cor.1, p.418.
74. Darwin, Charles, op cit., p.373.
75. Darwin to Henslow, 24 July–7 November 1834, Cor.1, pp.398, 400.
76. Darwin, Charles, op cit., p.399.
77. Ibid, pp.400–1.
78. Catherine Darwin to Darwin, 28 January 1835, Cor.1, p.424.
79. Darwin, Charles, op cit., p.407.
80. Ibid, p.413.
81. Darwin to Caroline, 10–13 March 1835, Cor.1, p.434.
82. FitzRoy, Captain Robert, op cit., p.229.
83. Darwin, Charles, op cit., p.418.
84. FitzRoy, Captain Robert, op cit., p.231.
85. Darwin, Charles, op cit., pp.422–3.
86. Ibid, p.424.
87. Ibid, pp.424–5.
88. Ibid, p.429.
89. Ibid, p.430.
90. Ibid, p.449.
91. Ibid, p.451.
92. Ibid, p.451.
93. Ibid, p.459.
94. Darwin to J. S. Henslow, 18 April 1835, Cor.1, p.440.
95. Darwin to Susan Darwin, 23 April 1835, Cor.1, pp.445-46.
96. Oxford Dictionaries Online.it.
97. Darwin, Charles, op cit., p.461.
98. FitzRoy, Captain Robert, op cit., p.258.
99. Darwin, Charles, op cit., p.494.
100. Ibid, p.497.
101. Darwin to Caroline Darwin, July/August 1835, Cor.1, p.458.
102. Darwin to W. D. Fox, 9–12 August 1835, Cor.1, pp.460-6.
103. Darwin to H. S. Fox, 15 August 1835, Cor.1, p.463.
104. Darwin to Alexander Burns Usborne, c. 1–5 September 1835, Cor.1, p.464.

CHAPTER 9

1. Darwin, Charles, op. cit., p.509.
2. Ibid, p.516.
3. Ibid,p.535.
4. Ibid, pp.537–8.
5. Ibid, p.538.
6. Ibid, p.539.

7. Ibid, p.541.
8. Ibid, p.549.
9. Ibid, p.550.
10. Ibid, p.554.
11. FitzRoy, Captain Robert, op. cit., p.298.
12. Darwin, Charles, op. cit., p.568.
13. Ibid, p.568.
14. Ibid, p.569.
15. Ibid, p.586.
16. Ibid, p.587.
17. Ibid, p.592.
18. Ibid, p.606.
19. Ibid, p.610.
20. Ibid, p.615.
21. Ibid, p.631.
22. Ibid, p.632.
23. Ibid, p.655.
24. Ibid, p.659.
25. Ibid, p.665.
26. Ibid, p.666.
27. Ibid, p.670.
28. Ibid, pp.675–6.
29. Ibid, p.684.
30. Ibid, p.677.

CHAPTER 10

1. Darwin, Francis, p.41.
2. Ibid, p.36.
3. Oxford Dictionaries Online.
4. Darwin, Francis, p.37.
5. Oxford Dictionaries Online.
6. Darwin, Francis, p.44.
7. Darwin, C. R., 1844, *Geological Observations on the Volcanic Islands visited during the Voyage of* HMS *Beagle, etc.* p.74
8. Ibid, p.93 (Footnote).
9. Darwin, Francis, p.39.
10. Ibid, pp.39, 41.
11. Darwin to Josiah Wedgwood (II), 5 October 1836, Litchfield, Henrietta. *Emma Darwin: A Century of Family Letters 1792–1896*, Volume 1, p.271.
12. FitzRoy, Captain Robert, p.359.
13. Darwin, Francis, p.46.
14. Darwin to J. S. Henslow [30/31 October 1836], Cor.1, p.512.
15. 'U.K. Scientists find Lost Darwin Fossils' 17 January 2012. http://www.technogypsie. com/science/?tag=dr-howard-falcon-lang and 'Why Evolution is True: Cache of Darwin Fossils Found'. http://whyevolutionistrue. wordpress.com/2012/01/18/cache-of-darwins-fossils-found
16. Emma Wedgwood to Mrs Hensleigh Wedgwood, October 1836, Litchfield, Henrietta. *Emma Darwin: A Century of Family Letters 1792–1896*, Volume 1, p.272, and Note (i).
17. Emma Wedgwood to Mrs Hensleigh Wedgwood, 21 November 1836, Litchfield, Henrietta. *Emma Darwin: A Century of Family Letters 1792–1896*, Volume 1, p.273.
18. Litchfield, Henrietta, *Emma Darwin: A Century of Family Letters 1792-1896*, Volume 1, p.149.
19. Emma Darwin's Diaries 1824–96: Dr John van Wyhe, editor, 2002, *The Complete Works of Charles Darwin Online*. (http://darwin-online.org.uk/)
20. Litchfield, Henrietta, op. cit., p.141.

21. Ibid, p.210.
22. Ibid, p.250.
23. Darwin to Charles Lyell, 13 February 1837, Cor.2, p.4. 24.
24. Darwin, Francis, p.42.
25. Darwin to Caroline, 19 May–16 June 1837, Cor.2, p.19.
26. Darwin to William Whewell, 18 June 1837, Cor.2, p.24.
27. Darwin, Francis, pp.56–7.
28. Darwin to John Richardson [24 and 25 July 1837], Darwin Correspondence Project, Letter 366f.
29. Darwin, Francis, pp.44–5.
30. Darwin to Susan Darwin, 1 April 1838, Cor.2, p.80.
31. Darwin to Charles Lyell, 9 August [1838], Cor.2, p.97
32. *The Correspondence of Charles Darwin: Volume 4, 1847–1850*, p.383.
33. Cor.2, pp.443-5.
34. Darwin, Francis, p.57.

CHAPTER 11

1. Malthus, *An Essay on the Principle of Population*, and *A Summary view of the Principle of Population*, p.217.
2. Ibid, p.70.
3. Malthus, op. cit., p.71.
4. Ibid, pp.158–9.
5. Ibid, p.159.
6. Ibid, p.160.
7. Ibid, p.225
8. Ibid, pp.231–2.
9. Ibid, p.236.
10. Ibid, p.238.
11. Ibid, pp.242–3.
12. Ibid, pp.250, 253.
13. Ibid, p.251.
14. Ibid, p.271.
15. Oxford Dictionaries Online.
16. Malthus, op. cit., p.272.
17. Oxford Dictionaries Online.
18. Darwin, Francis, op. cit., p.57.
19. Oxford Dictionaries Online.
20. Darwin, Francis, op. cit., p.57.
21. Ibid.
22. Ibid, pp.57–8.
23. Ibid.

CHAPTER 12

1. Darwin to John Stevens Henslow, 3 November 1838, Darwin Correspondence Project, Letter 429a.
2. Litchfield, Henrietta, *Emma Darwin: A Century of Family Letters 1792-1896*, Volume II, p.1. Darwin to Lyell, [12 November 1838].
3. Emma to Madame Sismondi, 15 November 1838, Litchfield, Henrietta. *Emma Darwin: A Century of Family Letters 1792-1896*, Volume II, p.6.
4. Darwin: 'Questions for Mr Wynne', February–July 1838. Cor.2, p.70.
5. Darwin to Emma, 2 January 1839, Litchfield, Henrietta. *Emma Darwin: A Century of Family Letters 1792–1896*, Volume II, pp.20–1.
6. Darwin to Emma, 20 January 1839, Litchfield, op. cit., p.25.
7. Ibid, Volume II, pp.31–2. Emma to Elizabeth Wedgwood, 5 February 1839.

8. Elizabeth Wedgwood to Madame Sismondi, 5 June 1839, Litchfield, op. cit., p.43.
9. Darwin to W. D. Fox, 24 October 1839, Cor.2, p.234.
10. Darwin, Francis. *Autobiography of Charles Darwin*, p.43.
11. Darwin to *Gardeners' Chronicle*, 16 August 1841, Cor.2, p.300.
12. Reeve, Tori, *Down House: the Home of Charles Darwin*, p.27
13. Ibid, p.6.
14. Darwin to Robert FitzRoy, 31 March 1843, Cor.2, p.354.
15. Oxford Dictionaries Online.
16. Darwin to G. R. Waterhouse, 26 July 1843, Cor.2, pp.375–6.
17. Darwin to Joseph Dalton Hooker, 11 January 1844, *The Correspondence of Charles Darwin: Volume 3, 1844-1846*, p.2.
18. Oxford Dictionaries Online.
19. Darwin to Leonard Jenyns, 12 October 1844, Cor.3, p.67.
20. Darwin to J. D. Hooker [10 September 1845], Darwin Correspondence Project, Letter 915.
21. Darwin to J. S. Henslow, 28 October 1845, Cor.3, p.260
22. Darwin to S.E. Darwin, 3 September 1845, Darwin Correspondence Project, Letter 913.
23. Cor.4, p.167.
24. Darwin to J. D. Hooker, 28 March [1849], Darwin, Francis (editor), *The Life of Charles Darwin*, p.174.
25. Darwin to J. D. Hooker, 13 June [1849], Darwin, Francis. *The Life and Letters of Charles Darwin, Including an Autobiographical Chapter*, Volume II, p.356.
26. Darwin to J. D. Hooker, 25 September [1853], Darwin, Francis, *The Life and Letters of Charles Darwin, Including an Autobiographical Chapter*, Volume II, p.359.
27. Oxford Dictionaries Online.
28. Darwin, Francis, *Autobiography of Charles Darwin*, pp.55-6.
29. *Genesis*, 1:1-27.
30. Darwin to W. D. Fox, 19 March [1855], *The Correspondence of Charles Darwin: Volume 5, 1851-1855*, p.288.
31. Cor.5, p.346, note 4.
32. J. D. Hooker to Darwin, 6–9 June 1855, Cor.5, p.345.

CHAPTER 13

1. Wallace, Alfred Russel, *My Life: A Record of Events and Opinions*, p.59.
2. Ibid, p.61.
3. Ibid, p.123
4. Ibid, p.127.
5. Ibid, p.124.
6. Ibid, p.144.
7. Chambers, Robert, *Vestiges of the Natural History of Creation*, p.24.
8. Ibid, p.25.
9. Ibid, p.54.
10. Ibid, pp.109–10.
11. Ibid, p.125.
12. Ibid, p.132.
13. Ibid, p.148.
14. Wallace, op. cit., p.144.
15. Ibid,
16. Oxford Dictionaries Online.
17. Wallace, op. cit., p.144
18. Ibid, p.149.
19. Ibid, p.147.
20. Ibid, pp.155–6.
21. Ibid, p.164.
22. Ibid, p.183.

23. Darwin to J. D. Hooker, 9 May [1856], C.6. p.106.
24. Darwin, Francis, p.58.
25. Wallace, op. cit., p.183.
26. Darwin to A. R. Wallace, 22 December 1857, *The Correspondence of Charles Darwin: Volume 6, 1856–1857*, pp.514–15.
27. Darwin to A. R. Wallace, 1 May 1857, Darwin Correspondence Project, Letter 2086.
28. Wallace, op. cit., p.184.
29. Darwin, Francis, op. cit., p.58.
30. Darwin to Charles Lyell, 18 [June 1858], *The Correspondence of Charles Dar win: Volume 7, 1858–1859,* p.107.
31. Wallace, op. cit., p.189.
32. Ibid, pp.189–90.
33. Ibid, p.190.
34. Ibid, p.191.
35. Darwin to Charles Lyell [25 June 1858], Cor.7, p.118.
36. J. D. Hooker and Charles Lyell to the Linnaean Society, 3 June 1858, Cor.7, pp.122–3.
37. Cor.7, p.130, Note 3.
38. Darwin to J. D. Hooker, 13 [July 1858], Cor.7, p.129.
39. *Journal of the Proceedings of the Linnean Society*, *Zoology* 3, 30 August 1858, pp.45–62.
40. Darwin, Francis, op. cit., pp.57–8.
41. Ibid, p.59.
42. Darwin to A. R. Wallace, 6 April 1859, Cor.7, p.279

CHAPTER 14

1. Darwin to J. S. Henslow, 10 November [1855], Cor.5, p.500.
2. Darwin to Edgar Leopold Layard, 9 December 1855, Cor.5, p.524.
3. Darwin to Henslow, 22 January [1856], Cor.6, p.25.
4. Cor.6, pp.55 and 56, note 2.
5. Darwin to Laurence Edmondston, 3 May [1856], Cor.6, p.99.
6. Darwin to W. D. Fox, 8 June 1856, Cor.6, p.135.
7. Darwin to T. V. Wollaston, 6 June 1856, Cor.6, p.134.
8. Wollaston, T. Vernon, *On the Variation of Species, with Special Reference to the Insecta; followed by an Enquiry into the Nature of Genera*, pp.180, 188.
9. Darwin to J. D. Hooker, 17/18 June 1856, Cor.6, p.147.
10. Darwin to Charles Lyell, 5 July 1856, Cor.6, p.169.
11. Darwin to Charles Lyell, 10 November 1856, Darwin Correspondence Project, Letter 1984.
12. Darwin to S. P. Woodward, 18 July 1856, Cor.6, p.189.
13. *Hamlet*, Act 3, Scene 1.
14. Darwin to J. D. Hooker, 30 July 1856, Cor.6, p.193.
15. Darwin to Frances E. E. Wedgwood, 18 August 1856, Cor.6, p.205.
16. Darwin to J. D. Hooker (after 20 January 1857), Cor.6, p.325.
17. Darwin to J. D. Dana. 5 April 1857, Cor.6, p.367 (Owen 1857b).
18. Darwin to Charles Lyell, 13 April 1857, Cor.6, p.376.
19. Forbes, Edward, Essay 'On the Connexion Between the Distribution of the Existing Fauna and Flora of the British Isles', p.399.
20. Ibid, p.350.
21. Darwin to J. D. Hooker, 3 June 1857, Cor.6, p.407.
22. Darwin to T. C. Eyton, 26 June 1857, Cor.6, p.417.
23. Burkhardt, Frederick (editor), *Charles Darwin's Letters: A Selection 1825–1859*, New York: Cambridge University Press, 1996, pp.177–9.
24. Darwin to Asa Gray, 29 November 1857, Cor.6, p.492.
25. Darwin to William Erasmus Darwin, 11 February 1858, *The Correspondence of Charles Darwin: Volume 7,*

1858–1859, Cor.7, p.21.
26. Darwin to J. D. Hooker, 23 June 1858, Cor.7, p.115.
27. Darwin to W. E. Darwin, 22 September 1858, Cor.7, p.158.
28. Darwin to Huxley, 18 January 1861, C9, p.1 and p.2 note 9.
29. Darwin to W. D. Fox, 12 February 1859, Cor.7, p.247.
30. Darwin to George Howard Darwin, 24 February 1859, Cor.7, pp.251–2.
31. Darwin to Charles Lyell, 20 October 1859, Cor.7, p.354.
32. Richard Dawkins presents *The Genius of Charles Darwin*, C. IWC Media Ltd, Channel 4 DVD, 2008. Professor Randolph M. Nesse, in conversation with Professor Richard Dawkins.
33. Cor.7, pp.371, 373.
34. *The Correspondence of Charles Darwin: Volume 12, 1864*, p.509.
35. Darwin, Francis, op. cit., p.59.

CHAPTER 15

1. Darwin, Francis, op. cit., p.622.
2. Darwin, Charles, *The Origin of Species*, p.x.
3. Ibid, p.27.
4. Ibid, pp.41–2.
5. Ibid, p.55.
6. Ibid, p.57.
7. Ibid, p.59.
8. Ibid, p.72.
9. Ibid, pp.78–9.
10. Ibid, p.111.
11. Ibid, p.122.
12. Ibid, p.131.
13. Ibid, p.194.
14. Ibid, p.224.
15. Ibid, p.256.
16. Ibid, p.287.
17. Ibid, p.293.
18. Ibid, p.317.
19. Ibid, p.318.
20. Ibid, p.320.
21. Ibid, p.343.
22. Ibid, p.346.
23. Ibid, pp.348–9.
24. Ibid, p.351.
25. Ibid, p.350.
26. Ibid, p.423.
27. Ibid, p.442.
28. Ibid, p.443.
29. Ibid, p.444.
30. Ibid, p.446.
31. Ibid, p.450
32. Ibid, p.450.
33. Darwin to Ernst Haeckel, [after 10] August – 8 October 1864, Cor.12, p.302.
34. Darwin, Francis, op. cit., p.57.
35. Darwin, Charles, op. cit., p.30.
36. Darwin to J. L. A. de Quatrefages de Bréau, 5 December 1859, Cor.7, p.415.
37. Darwin to Francis Galton, 9 December 1859, Cor.7, p.417.
38. Darwin to Asa Gray, 21 December 1859, Cor.7, p.440.

39. Darwin to Adam Sedgwick, 24 November 1859, Cor.7, p.396.
40. Darwin to Adam Sedgwick, 26 November 1859, Cor.7, p.404.
41. Latin Vulgate Bible, 3 Esdras. 4:41.
42. Darwin, Francis, op cit., p.66.
43. Ibid, pp.66–7.
44. Ibid, p.59.

CHAPTER 16

1. Darwin to Thomas Bridges *The Correspondence of Charles Darwin: Volume 8, 1860,* 6 January 1860, p.19, and p.20, note 1.
2. Darwin to Baden Powell, 18 January 1860, Cor.8, p.40.
3. Darwin to Charles Lyell, Cor.8, 15 April 1860, p.161.
4. Darwin to J. S. Henslow, 8 May, Cor.8, p.195.
5. Darwin to J. S. Henslow, 14 May, Cor.8, p.208.
6. Darwin to A. R. Wallace, 18 May 1860, Cor.8, p.220.
7. Darwin to Asa Gray, 22 May, Cor.8, p.223.
8. Darwin to Asa Gray, 3 July, Darwin Correspondence Project, Letter 2855.
9. These reports, published by the *Athenaeum*, were dated 7 July and 14 July 1860.
10. *Athenaeum*, 7 July 1860, pp.25–6, Cor.8, pp.591–3.
11. Ibid, 14 July 1860, pp. 64–5, Cor.8, pp.593–7.

CHAPTER 17

1. Darwin to Asa Gray, 22 July, Cor.8, p.299.
2. Darwin to John Murray, 3 August, Cor.8, p.309.
3. Darwin to Hooker, 23, Cor.9, p.100.
4. Darwin to Hooker, 25 January, Cor.10, p.48.
5. Darwin to Armand de Quatrefages, 11 July, Cor.10, pp.313–14.
6. Cor.11, pp.754–65.
7. Darwin to *Athenaeum*, 18 April, Cor.11, p.324.
8. Darwin to George Bentham, 22 May, Cor.11, pp.432–3.
9. Darwin to George Bentham, 19 June, Cor.11, p.497.
10. Darwin to Henry Fawcett, 6 December [1860], Cor.18, p.375.
11. Wallace, Alfred Russel, op. cit., p.197.
12. Darwin to Hooker, 4 February, Cor.9, p.20.
13. Darwin to Jean L.A. Quatrefages de Bréau, 25 April, Cor.9, p.102.
14. Darwin to W. E. Darwin, 9 May, Cor.9, p.123.
15. Darwin to Hooker, 18, Darwin Correspondence Project, Letter 3152.
16. Darwin to John S. W. Kershaw, 23 May, Cor.9, p.135.
17. Darwin to Bartholomew J. Sulivan, 24 May, Cor.9, p.138.
18. Darwin to Frances Julia Wedgwood, 11 July, Cor.9, p.200.
19. Darwin to Asa Gray, 17 September, Cor.9, p.267.
20. Darwin to Asa Gray (after 11 October 1861), Cor.9, p.302.
21. Darwin to Wallace, 30 November 1861, Cor.9, p.357.
22. Darwin to Hooker, Cor.9, p.367.
23. Darwin to Asa Gray, 22 January, Cor.10, p.40.
24. Oxford Dictionaries Online.
25. Darwin to Charles Kingsley, 6 February, Cor.10, pp.71-2.
26. Darwin to Hooker, 18 March, Cor.10, p.123.
27. Cor.10, p.146.
28. Cor.6, p.515, note 12.
29. Darwin to Asa Gray, 10–20 June, Cor.10, p.241.

30. Emma to T.G. Appleton, 28 June, Cor.10, p.275.
31. Darwin to Hooker 24 December, Cor.10, p.625.
32. Darwin to James Dwight Dana, 7 January, Cor.11, p.23.
33. Darwin to James Dwight Dana, 7 January, Cor.11, p.23.
34. Darwin to Hugh Falconer, 28 January, Cor.11, pp.61–2.
35. Hugh Falconer to Darwin, 18 January, Cor.11, p.55.
36. Darwin to H. W. Bates, 26 January, Cor.11, p.83.
37. Cor.11, p.119.
38. Darwin to Robert Scot Skirving, 16 November, Cor.18, p.383.
39. Darwin to J.D. Hooker, 13, Cor.11, p.225.
40. Darwin to Asa Gray, 20 March, Cor.11, p.247.
41. Darwin to John Scott, 2 July, Cor.11, p.519.
42. Cor.12, p.455, note 3.
43. Darwin to Charles Lyell, 22 January, Cor.11, p.35.
44. Darwin to Hooker, 9 February, Cor.13, p.56.
45. Darwin to Hooker, 27, Cor.13, pp.245–6.
46. Darwin to Hooker, 22 December, Cor.13, p.329.
47. Cor.14, pp.1–2.
48. Darwin to Hooker, Cor.14, p.84.
49. Litchfield, op. cit., Volume II, pp.184–5.
50. Darwin to A. R. Wallace, 5 July [1866], Cor.14, pp.235–6.
51. Spencer, Herbert, *Principles of Biology*, 1864–7. I, pp.444–5.

CHAPTER 18

1. Wallace, *My Life; a Record of Events and Opinions*, p.115.
2. Ibid, p.201.
3. Ibid, p.205.
4. Ibid, p.208.
5. Wallace to Hooker. 6 October [1858], Cor.7, p.166.
6. Darwin to Wallace, 20 April [1870], Cor.18, pp.100–1.
7. Information kindly supplied by Darwin Correspondence Project.
8. Wallace, op. cit., p.245.
9. Ibid, p.247.
10. Ibid, p.178.
11. Ibid, p.201.
12. Oxford Dictionaries Online.
13. Wallace, op. cit., p.327.
14. Ibid, p.118.
15. Ibid, p.11.
16. Ibid, p.45.
17. Ibid, p.46.
18. Ibid, p.119.
19. Ibid, p.119.
20. Oxford Dictionaries Online.
21. Ibid.
22. Ibid.
23. Wallace, op. cit., p.334.
24. Ibid, p.335.

25. Oxford Dictionaries Online.
26. Wallace, op. cit., pp.340, 345–6.
27. Ibid, p.336.
28. Ibid, p.338.
29. Ibid, p.355.
30. Ibid, p.380.
31. Ibid, p.252.
32. Oxford Dictionaries Online.
33. Wallace, op. cit., pp.239–40.
34. Darwin, Francis, op. cit., Volume III, pp.368–9.
35. Wallace, op. cit., p.381.
36. Ibid, pp.236–7.

CHAPTER 19

1. Darwin, Charles, *The Variation of Animals and Plants under Domestication*, Volume I, p.1.
2. Ibid, Volume I, pp.8–9.
3. Ibid, Volume I, p.8.
4. Ibid, Volume II, p.271.
5. Oxford Dictionaries Online.
6. Darwin, *Variation*, Volume II, p.319.
7. Ibid, Volume II, p.509.
8. Ibid, Volume II, p.478.
9. Ibid, Volume I, pp.343–4.
10. Darwin to Wallace, 22 November, Cor.18, p.303.
11. Ibid, Volume I, p.353.
12. Oxford Dictionaries Online.
13. Darwin, *Variation*, Volume I, p.360.
14. Ibid, Volume II, pp.484, 509.
15. Darwin to St G.J. Mivert, Cor.19, p.34.
16. Darwin, *Variation*, Volume II, p.199.
17. Oxford Dictionaries Online.
18. Darwin, *Variation*, Volume II, op. cit., p.70.
19. Ibid, Volume II, p.432.
20. Ibid, p.3.
21. Ibid, p.504.
22. Ibid, Volume II, pp.523–4.
23. Ibid, Volume II, p.508.
24. Ibid, Volume I, p.2.
25. Ibid, Volume I, p.3.
26. Ibid, Volume II, p.285.
27. Ibid, Volume II, p.278.
28. Ibid, Volume II, p.288.
29. Ibid, Volume II, p.282.
30. Ibid, Volume I, p.349.
31. Ibid, Volume II, p.457.

NOTES

32. Simpson, J. A. and E. S. C. Weiner, *The Oxford English Dictionary*.
33. Darwin, *Variation*, Volume II, op. cit., p.433.
34. Ibid, Volume II, p.461.
35. Ibid, Volume II, p.459.
36. Darwin, *Variation*, Volume II, p.471.
37. Ibid, Volume II, p.320.
38. Ibid, Volume II, p.464.
39. Ibid, Volume II, p.484.
40. Ibid, Volume II, p.489.
41. Ibid, Volume II, p.491.
42. Ibid, Volume II, p.491.
43. Darwin to William Ogle, 6 March, *The Correspondence of Charles Darwin: Volume 16, 1868,* pp.244–5.
44. Orel, *Heredity before Mendel*
45. Darwin to Carus, 21 March, Cor.16, p.288.
46. Darwin to H. W. Bates, 18 March, Cor.16, p.279.
47. Darwin to J. J. Weir, 18 April, Cor.16, p.413.
48. J. D. Hooker to Darwin, 16 June 1868, Cor.16, p.583.
49. Darwin to J. D. Hooker, 17, Cor.16, p.584.
50. A. R. Wallace to Darwin, 30 August, Cor.16, p.705.
51. Darwin to James Orton, 7 October, *The Correspondence of Charles Darwin: Volume 17, 1869,* p.420.

CHAPTER 20

1. Galton, Francis, *Hereditary Genius*, p.211.
2. Ibid, p.71.
3. Ibid, p.114.
4. Ibid, p.151.
5. Ibid, p.158.
6. Ibid, p.165.
7. Ibid, p.31.
8. Ibid, p.68.
9. Ibid, pp.224–5.
10. Ibid, p.227.
11. Ibid, p.228.
12. Ibid, pp.228–9.
13. Ibid, p.230.
14. Ibid, p.231
15. Ibid, p.236.
16. Oxford Dictionaries Online.
17. Darwin to Francis Galton, 23 December, Cor.17, pp.530-1.
18. Francis Galton to Darwin, 24 December 1869, Cor.17, p.532.
19. 1 Corinthians 13:12.

NOTES

CHAPTER 21

1. Darwin to Wallace, 22 December 1857, Cor.6, pp.514–15. In Darwin, *The Descent of Man,* p.xxxi.
2. Darwin, *Descent*, p.18.
3. Oxford Dictionaries Online.
4. Darwin, *Descent,* pp.42–3.
5. Ibid, p.44.
6. Ibid, p.46.
7. Ibid, p.65.
8. Ibid, p.65.
9. Ibid, p.96.
10. Ibid, pp.141–2.
11. Ibid, p.147.
12. Darwin, *Descent,* p.147.
13. Ibid, p.151.
14. Luke's Gospel 6:31.
15. Darwin, *Descent,* p.151.
16. Ibid, p.159.
17. Emma to Francis Darwin, 1885, in Litchfield, Henrietta, *Emma Darwin*, Cambridge University Press, 1904.
18. Darwin, *Descent*, p.153.
19. Ibid, pp.159–60.
20. Ibid, pp.163–4.
21. Ibid, p.168.
22. Ibid, p.176.
23. Genesis I, 26–7.
24. Darwin, *Descent,* p.207.
25. Ibid, p.213.
26. Oxford Dictionaries Online.
27. Darwin, *Descent*, p.245.
28. Darwin, *Descent,* p.247.
29. Ibid, p.618.
30. Darwin, *Descent,* p.677.
31. Ibid, p.682.
32. Ibid, p.688.
33. Ibid, pp.688–9.
34. Ibid, p.689.
35. Ibid.
36. Source, Smithsonian Institution, National Museum of Natural History, Washington. DC.
37. Darwin to F.J. Wedgwood, Cor.19, p.247.
38. Oxford Dictionaries Online.
39. Darwin to Henry Holland, Cor.19, p.642.
40. Darwin to Horace Darwin, 15 December 1871, *Emma Darwin*: *A Century of Family Letters 1792–1896*, Volume 2, p.207.
41. Cor.1, p.290.

NOTES

CHAPTER 22

1. Darwin to Vernon Lushington, Cor.17, p.127.
2. Oxford Dictionaries Online.
3. Ibid.
4. Ibid.
5. Ibid.
6. Morgan, *The Oxford Illustrated History of Britain*, p.351.
7. *The Royal Society: 350 years of Science*, pp.14–15.
8. Ibid, p.10.
9. Ibid, p.19.
10. Ibid, p.35.
11. Ibid, p.39.
12. Ibid, p.48.
13. Tritton, *The Hutchinson Encyclopedia for the Millennium*.

CHAPTER 23

1. Matthew, and Harrison, *Oxford Dictionary of National Biography*.
2. Darwin, Erasmus, *Zoonomia; or The Laws of Organic Life*, Vol. I.
3 *Book of Common Prayer*, Collins, London and Glasgow.
4. Darwin, Charles, *The Life of Erasmus Darwin*, p.37.
5. Ibid, pp.84–9.
6. Ibid, p.35.
7. Ibid, p.57.
8. Ibid, p.59–60.
9. Ibid.
10. Oxford Dictionaries Online.
11. Darwin, Charles, *The Life of Erasmus Darwin*, p.63.
12. Darwin, Francis, op. cit., p.13.

CHAPTER 24

1. Lamarck, *Zoological Philosophy*, p.xvii.
2. Ibid, pp.xvii–xx.
3. Ibid, p.36.
4. Ibid, pp.1–2.
5. Ibid, p.2.
6. Ibid, p.122.
7. Ibid, p.242.
8. Ibid, p.247.
9. Ibid, pp.178.
10. Ibid, pp.179–80.
11. Ibid, p.36.
12. Ibid, pp.179–80.

NOTES

CHAPTER 25

1. In fact, Matthew matriculated in 1804/5 as a medical student, but did not graduate. Information kindly supplied by Edinburgh University Library, Special Collection.
2. *Quarterly Journal of Agriculture*, Volume III, February 1831–September 1832.
3. 'Patrick Matthew of Gourdiehill, Naturalist' by Scottish zoologist William T. Calman, published in 1912 in the British Association's *Handbook and Guide to Dundee and District*. A. W. Paton and A.H. Millar (editors). For his article, Calman relied upon information given to him by Patrick Matthew's daughter 'Miss Euphemia Matthew of Newburgh [Fife]'.
4. Matthew, *On Naval Timber and Arboriculture*, Appendix, Note A, p.363.
5. Letter from Darwin *to Gardeners' Chronicle*, published 21 April 1860, in Francis Darwin, op. cit., Volume II, p.95.
6. Darwin, Charles, *Origin*, pp.62–3.
7. Darwin, Francis, op. cit., Part II, p.96.
8. Darwin, Charles, *Origin*, p.20.
9. *Quarterly Review*, vol. xix, April–July 1833, p.125.
10. Loudon, John Claudius (1783–1834), was a designer of gardens and cemeteries, author of *The Encyclopaedia of Gardening* (1822) and of *The Encyclopaedia of Trees and Shrubs* (1842), and founder of the *Magazine of Natural History* (1828).
11. The review appeared under the heading '*Matthew, Patrick*: On Naval Timber and Arboriculture; with Critical Notes on Authors who have recently treated the Subject of Planting'.
12. *Gardeners' Magazine*, Volume VIII (1832), pp.703–4.
13. *United Services Journal and Naval and Military Magazine*, 1831 Part II, p.457.
14. *Gardeners' Magazine*, Volume VIII, p.208.
15. Information kindly supplied by Department of Manuscripts and University Archives, Cambridge University Library.
16. Information kindly supplied by Department of Manuscripts and University Archives, Cambridge University Library.
17. Matthew, op. cit., p.xi.
18. Ibid, pp.366–7.
19. Ibid, p.x.
20. Ibid, p.106.
21. Ibid, p.381.
22. Ibid, p.382.
23. Ibid, p.383.
24. Matthew, op. cit., p.384.
25. Ibid, p.385.
26. Ibid, p.x.
27. Darwin to Patrick Matthew, 13 June, Darwin Correspondence Project, Letter 3600.

CHAPTER 26

1. The full title was 'Two Essays: Upon a Single Vision with Two Eyes, the Other on Dew', published by Constable of London.
2. Darwin, *Origin*, op. cit., pp.18–19.
3. Ibid.

4. Darwin to Hooker, October 1865 in Francis Darwin, op. cit., Part II, p.225.
5. Information kindly supplied by the Centre for Research Collections, University of Edinburgh.
6. Wells, 'Two Essays: Upon a Single Vision with Two Eyes, the Other on Dew', pp.425–39.

CHAPTER 27

1. Darwin, Francis, op. cit., pp.38–9.
2. FitzRoy, op. cit., Volume II, p.76.
3. Darwin to Henslow, 14 October 1837, Cor.2, pp.51-2.
4. Darwin, Francis, op. cit., pp.43–4.
5. Ibid, p.45.
6. Ibid, p.53.
7. Darwin to Emma, Darwin Correspondence Project, Letter 704.
8. Darwin to Adolf von Morlot, 9 August 1844, Cor.3, p.51.
9. Darwin to Emma, 3–4 February 1845, Cor.3, p.132.
10. Darwin to Hooker, 31 March 1845, Cor.3, p.166.
11. Darwin to Susan, 3–4 September 1845, Cor.3, p.247.
12. Darwin to Hooker, 5 or 12 November 1845, Cor.3, p.264.
13. Darwin to Emma, 25 June 1846, Cor.3, p.326.
14. Darwin to Hooker, 8 or 15 July 1846, Cor.3, p.327.
15. Darwin to John Maurice Herbert, 3 September? 1846, Cor.3, p.338.
16. Darwin to Hooker, 7 April 1847, Cor.4, p.29.
17. Darwin to Bernhard Studer, 4 July 1847, Cor.4, p.54.
18. Darwin to Emma, 31 October 1847, Cor.4, pp.91-2.
19. Darwin to Emma, 27–8 May 1848, Cor.4, p.147.
20. Darwin, Francis. *Autobiography of Charles Darwin*, p.55.
21. Darwin Hooker, 28 March 1849, Cor.4, p.227.
22. Darwin to Richard Owen, 24 February 1849, Cor.4, p.219.
23. Gully, *The Water-cure in Chronic Diseases*, pp.328–74.
24. Darwin to Hooker, 28 March 1849, Cor.4, p.227.
25. Reeve, Tori, op. cit., p.17.

CHAPTER 28

1. Darwin to John Wickham Flower, 23 March, Cor.5, p.8.
2. Darwin to Fox, 24 October, Cor.5, p.100.
3. Oxford Dictionaries Online.
4. Darwin to Fox, 17 July, Cor.5, p.147.
5. Down House MS, reproduced in Colp, Ralph, *Darwin's Illness*, pp.187–257.
6. Cor.5, p.190.
7. Darwin to Hooker, 14 August, Cor.5, p.408.
8. Darwin, Francis, op. cit., pp.54–5.
9. Cor.10, p.190, pp.xxii–xxiii.
10. Colp, *Darwin's Illness*, p.170.
11. Darwin to Hooker, Cor.5, p.507.
12. Darwin to Fox, 3 October, Cor.6, p.238.

13. Darwin to Lyell, 13 April, Cor.6, p.376 and p.377, note 5.
14. Darwin to Darwin, 13 May, Cor.6, pp.394-95.
15. Darwin to Hooker, 2 June, Cor.6, p.404.
16. Darwin to Hooker, 10 April, Cor.7, p.65.
17. Darwin to Lyell, 26 April, Cor.7, p.83.
18. Oxford Dictionaries Online.
19. Darwin to Emma, Cor.7, p.84.
20. Darwin to Fox, 8 May, Cor.7, p.90.
21. Darwin to Hooker, 8, Cor.7, p.102.
22. Darwin to Syms Covington, 16 January 1859, Cor.7, p.235.
23. Darwin to John Phillips, 8 February, Cor.7, pp.245-6.
24. Darwin to Fox, 12, Cor.7, p.247
25. Darwin to Hooker, 18 May 1859, Cor.7, p.299.
26. Darwin to Hooker, Cor.7, p.300.
27. Cor.7, p.343.
28. Darwin to Fox, Cor.7, p.377.
29. Darwin to Fox, 25 December, Cor.7, p.449.
30. Darwin to Hooker, 22 January 1860, Cor.8, p.45 and note 4.
31. Darwin to Fox, 22, Cor.8, p.133.
32. Darwin to Lyell, 25, Cor.8, p.265.
33. Darwin to Hooker, 23, Cor.9, pp.98-9.
34. Darwin to W.B. Tegetmeier, 29 August, Cor.9, p.242.
35. Darwin to Hooker, Cor.10, p.419.
36. Darwin to Hugh Falconer, 5 January, Cor.11, p.11.
37. Darwin to Fox, 23 March, Cor.11, p.255.
38. Darwin to Hooker, 26, Cor.11, p.265.
39. Cor.11, p.603, note 6 and p.655 note 3.
40. Cor.11, p.643, note 5.
41. Darwin to Hooker, Cor.12, p.31.
42. Darwin to Gray, 25 February, Cor.12, p.60.
43. Emma to Fox, Cor.12, pp.168-69.
44. Darwin to Buckland, 15 December, Cor.12, p.471.
45. Cor.13, p.147, note 3.
46. Darwin to John Chapman, 7 June 1865, Cor.13, p.179.
47. Cor.7, p.482.
48. Darwin to Gray, 15 August, Cor.13, p.223, and note 12.
49. Darwin to Edward Cresy, 7 September, Cor.13, pp.229-30.
50. Darwin to Hooker, 27, Cor.13, pp.245-46.
51. Darwin to Fox, 25/26 October, Cor.13, p.284.
52. Darwin to Bence Jones, 3 January, Cor.14, p.4.
53. Darwin to Hooker, 3 February, Cor.16, p.65.
54. Darwin to J.J. Moulinié, 22 February, Cor.17, p.93.
55. Darwin to W.C. Tait, 18 April, Cor.17, p.183.
56. Darwin to Hooker, 22 June, Cor.17, p.279.
57. Darwin to H.B. Jones, 3 August, Cor.18, p.224.
58. Darwin to W.S. Dallas, 27 January, Cor.19, p.44.

59. Darwin to Matthew, 15 March, Cor.19, p.182.
60. Darwin to F.E. Abbot, 16 November, Cor.19, p.686.
61. Hayman, Professor John A., 'Darwin's Illness Revisited', *British Medical Journal* 2009:339:b4968. 13 December 2009.
62. Berg, Tymoczko, Stryer, and Clarke, *Biochemistry*, p.442.
63. http://barryjmarshall.blogspot.com/2009/02/darwins-illness-was-helicobacter-pylori.html
64. Mandell, Bennett, and Dolin, *Principles and Practice of Infectious Diseases*, p.2286.
65. Ibid, p.2288.
66. Ibid, pp.2288–9.
67. Oxford Dictionaries Online.
68. Warrell, Cox, Firth, and Benz Jr, *Oxford Textbook of Medicine*, p.561.
69. Ibid, p.562.
70. Ibid, p.563.

CHAPTER 29

1. Colp, *Darwin's Illness*, p.156.
2. Oxford Dictionaries Online.
3. Darwin, Francis, op. cit., p.53.
4. Darwin to Hooker, 10 [November 1863], Cor.11, p.666.
5. Darwin to Charles Boner [before 8 January 1870], Cor.18, p.5.
6. Darwin, Francis, op. cit., p.53.
7. Adler, 'Darwin's Illness', p.1102.
8. Oxford Dictionaries Online.
9. Darwin to Hooker, 31 March 1845, Cor.11, p.166.
10. Adler, op. cit., p.1103.
11. Ibid, p.1103.
12. Ibid, p.1103.
13. Darwin, Charles, *The Voyage of the Beagle*, p.451.
14. Oxford Dictionaries Online.
15. Other vectors of Chagas' disease are *Triatoma dimidiata*, which is indigenous to Ecuador and Columbia, and *Rhodnius prolixus*, which is indigenous to Colombia, Venezuela, and Guyana.
16. Adler, op. cit., p.1103.
17. Oxford Dictionaries Online.
18. Adler, op. cit., p.1103.
19. 'Nova tripanosomiase humana. Estudos sobre ea morfologia e o ciclo evolutivo do *Schizotrypanum cruzi*, n. gen., n. sp., ajente etiologico de nova entidade morbida do homem'. *Memorias do Instituto Oswaldo Cruz*, 1, pp.159–218, in Cox, *The Wellcome Trust Illustrated History of Tropical Diseases*, pp.199–200.
20. Chagas, 'Nova tripanosomiase humana. Estudos sobre ea morfologia e o ciclo evolutivo do *Schizotrypanum cruzi*, n. gen., n. sp., ajente etiologico de nova entidade morbida do homem'. *Memorias do Instituto Oswaldo Cruz*, 1, pp.159–218, in Cox, op. cit., p.200.
21. Dias, 'Estudos sobre o Schizotrypanum cruzi'. *Memorias do Instituto Oswaldo Cruz*, 28, 1-110, in Cox, op. cit., p.202.
22. Cox, op. cit., p.193.
23. Tiexeira, Nascimento, and Sturm, 'Evolution and Pathology in Chagas' Disease – A Review, *Mem*

Inst Oswaldo Cruz, Rio de Janeiro, Vol. 101(5) August 2006, p.470.
24. http://en.wikipedia.org/wiki/Charles_Darwin's_health#The_Chagas_hypothesis
25. Oxford Dictionaries Online.
26. Tiexeira, Nascimento, and Sturm, op. cit., p.471.
27. Pan American Health Organization/World Health Organization. Source: PAHO/WHO, Program on Communicable Diseases. *However, this is not the case to the west or east of the Andes range, where Chagas' disease was (and is) endemic.*
28. Vazquez-Prokopec, Spillmann, Zaidenberg, Kitron, and Gürtler, 'Cost-Effectiveness of Chagas' Disease Vector Control Strategies in Northwestern Argentina', published online, 20 January 2009.
29. Tiexeira, Nascimento, and Sturm, op. cit., p.470.
30. Hayman, 'Darwin's Illness Revisited', *British Medical Journal* 2009:339:b4968. 13 December 2009.

CHAPTER 30

1. Emma to Darwin, 21–22 November 1838. Cor.2, p.123.
2. Gospel of St John, 13:34.
3. Emma to Darwin 23 January 1839, Cor.2, p.169.
4. Barlow, *The Autobiography of Charles Darwin 1809-1882* (with original omissions restored).
5. Emma to Darwin, Darwin, Emma. Op. cit., Volume II, p.172.
6. Image of letter, from Darwin Online, http://farm7.static.flickr.com/6070/6025646324_29000c6352_o.jpg
7. Darwin to Fox, 23 August 1841, Cor.2, p.303.
8. Darwin to Fox, 28 September 1841, Cor.2, p.305.
9. Statement by Henrietta Litchfield, née Darwin, Darwin, Emma, op. cit., Volume II, p.173.
10. Murphy, *Blanco White: Self-banished Spaniard*, p.21.
11. Ibid, p.24.
12. Ibid, p.57.
13. Ibid, p.78.
14. Ibid, p.193.
15. Emma to Madame Sismondi, 27 August 1845, Darwin, Emma, op. cit., Volume II, p.96.
16. Darwin, *Origin*, p.450.
17. Darwin to Hooker, 1 February, Darwin Correspondence Project, Letter 7471.
18. Darwin, Francis, op. cit., Volume III, pp.368–9.
19. Oxford Dictionaries Online.
20. Sagan, 'On the Origin of DNA', You Tube.
21. Darwin to Abbot, 6 September, Darwin Correspondence Project. Letter 7924.
22. Darwin to Gray, 22 May, Cor.8, pp.223-24.
23. Oxford Dictionaries Online.
24. Darwin to Hooker, 12 July, Cor.18, p.209.
25. Darwin to Lyell, 17 June, Cor.8, p.258.
26. Cor.8, p.259, note 2.
27. Darwin to Boole, 13 December 1866, Darwin Correspondence Project, Letter 5303.
28. Boole, *Collected Works*, Volume I, pp.vii–viii.
29. Cor.14, p.425, note 2.
30. Darwin to Boole, 14 December 1866, Cor.14, pp.425-6.

31. Darwin to Hooker, 8 February, Darwin Correspondence Project, Letter 5395.
32. Oxford Dictionaries Online.
33. Ibid.
34. Darwin, *Variations*, Volume II, p.526.
35. Ibid, p.525.
36. Darwin to Lubbock], Cor.7, p.388.
37. Isaiah 26:3.
38. Emma to Darwin, June 1861, Cor.9, p.155.
39. Darwin to Frances Julia Wedgwood, 11 July, Cor.18, pp.379-80.
40. Darwin to Vernon Lushington, Cor.17, p.127.
41. Darwin, Emma, op. cit., Volume II, p.196.
42. Darwin to Huxley, 21 September, Cor.19, p.591.
43. Darwin to Abbot, 16 November, Cor.19, p.686.

CHAPTER 31

1. Oxford Dictionaries Online.
2. AD is the abbreviation for Anno Domini – the year of Our Lord – i.e. the year of Christ's birth, and BC is the abbreviation for 'Before Christ'. However, scholars are uncertain about Christ's dates.
3. Gospel of St Matthew: 16, verse 16.
4. Gospel of St John: 14, verse 6.
5. Oxford Dictionaries Online.
6. American Psychiatric Association, *Diagnostic and Statistical Manual of Mental Disorders*, p.326.
7. Oxford Dictionaries Online.
8. Ibid.
9. Gelder, Harrison and Cowen. *Shorter Oxford Textbook of Psychiatry*, pp.7–8.
10. Darwin, *The Descent of Man*, p.682.
11. Oxford Dictionaries Online.
12. Darwin, op cit., p.117.

CHAPTER 32

1. Oxford Dictionaries Online.
2. Ibid.
3. Eagle, et al., 'Dinosaur Body Temperatures Determined from Isotopic (^{13}C-^{18}O) Ordering in Fossil Biominerals', *Science*, 23 June 2011.
4. Oxford Dictionaries Online.
5. Brattstrom, 'Body Temperatures of Reptiles', *American Midland Naturalist*. 1965. 73 (2): pp.376–422.
6. Price, Gregory D. and Elizabeth V Nunn, Valanginian Isotope Variation in Glendonites and Belemonites from Arctic Svalvard: Transient Glacial Temperatures during the Cretaceous Greenhouse', *Geology* 2010: 38. pp.251–4.
7. Oxford Dictionaries Online.
8. Alvarez, Alvarez, Asaro and Michel, 'Extraterrestrial Cause for the Cretaceous-Tertiary Extinction: Experiment and Theory', *Science* 208, 1980 (4448): 1095–1108.
9. Hildebrand, Penfield, Kring, Pilkington, Camargo, Jacobsen and Boynton, 'Chicxulub Crater; a

possible Cretaceous/Tertiary boundary impact crater on the Yucatan Peninsula, Mexico', September 1991. *Geology* 19 (9), pp.867–71.
10. Keller, Gerta, et al., 2004, 'Multiple impacts at the KT boundary and the death of the dinosaurs. Impact predates the K-T Boundary Mass Extinction', published in the *Proceedings of the Academy of Sciences of the United States of America*, 16 March 2004, Volume 101, Number 11, pp.3753–8.
11.Chatterjee, S. 1997, 'Multiple Impacts at the KT Boundary and the Death of the Dinosaurs', Proceedings of the 30th International Geological Congress 26: 31-54.
12. Keller, Sahni and Bajpai, 'Deccan Volcanism, the KT Mass Extinction and Dinosaurs', *Journal of Bioscience*, 34(5), November 2009.

CHAPTER 33

1. Oxford Dictionaries Online.
2. Huxley to Phillips, 31 December 1867, Oxford University Museum, 29, and Matthew and Harrison, *Oxford Dictionary of National Biography*.
3. Oxford Dictionaries Online.
4. Ibid.
5. 'Fossils Provide "Missing Link" between Dinosaur and Bird', by Richard Alleyne, Science Correspondent, 25 September 2009. *The Telegraph* (online).

CHAPTER 34

1. Oxford Dictionaries Online.
2. Malthus, *An Essay on the Principle of Population*, and *A Summary view of the Principle of Population*, pp.127–9.
3. Ibid, p.171.
4. Ibid, p.130
5. Ibid, p.172.
6. Darwin, Erasmus. 'The Temple of Nature', Notes, p.45, published posthumously in 1803, quoted in Charles Darwin's *The Life of Erasmus Darwin*, p.39.
7. Galton, *Inquiries into Human Faculty*, p.17, Footnote 2.
8. Ibid, p.2.
9. Ibid, pp. 6–7.
10. Ibid, pp.10–11.
11. Ibid, p.13.
12. Oxford Dictionaries Online.
13. Galton, op. cit., p.10.
14. Ibid, p.197.
15. Ibid, p.56.
16. Ibid, pp.198–9.
17. Ibid, p.199.
18. Ibid, pp.199–200.
19. Ibid, p.200–1.
20. Oxford Dictionaries Online.
21. Galton, op. cit., p.206.
22. Ibid, p.207.

23. Ibid, pp.218–19.
24. Ibid, pp.196–7.
25. Oxford Dictionaries Online.
26. Matthew and Harrison, *Oxford Dictionary of National Biography*.
27. Galton, 'Eugenics: its Definition, Scope, and Aims'. *American Journal of Sociology*, Volume X, July 1904; Number 1.
28. Matthew and Harrison, op. cit.,
29. Wallace, *My Life; a Record of Events and Opinions*, pp.299–300.
30. Ibid, p.386.
31. Chase, *The Legacy of Malthus: The Social Costs of the New Scientific Racism*, p.16, and note p.626.
32. Pearson, *National Life from the Standpoint of Science*, pp.20–2.
33. Ibid, p.20–1, 23.
34. Ibid, p.26.
35. Ibid, p.46
36. Ibid, p.64.

CHAPTER 35

1. Darwin, Major Leonard, 'First Steps Towards Eugenic Reform', *Eugenics Review*, Volume 4 (1), April 1912.
2. Darwin, Leonard, 'Eugenicists Hail Their Progress as Indicating Era of Supermen', *New York Herald Tribune* (1932), review of Third International Eugenics Congress.
3. Darwin, Leonard, *The Need for Eugenic Reform*, Dedication.
4. Oxford Dictionaries Online.
5. Darwin, Francis, op. cit., p.62.

CHAPTER 36

1. Oxford Dictionaries Online.
2. Arendt, *The Origins of Totalitarianism*, p.178.
3. Oxford Dictionaries Online.
4. Ibid.
5. Ibid.
6. Ibid.
7. Hitler, *Mein Kampf*, p.26.
8. Ibid, p.277.
9. Hamann, *Hitler's Vienna: a Dictator's Apprenticeship*, p.217.
10. Hitler, op. cit., p.41.
11. Daim, *Der Mann, der Hitler die Ideen gab*, p.34. See also Norman, *Hitler: Dictator orPuppet*, pp.51–4.
12. List, *Das Geheimnis der Runen*, 2a, p.81f. In Goodrick-Clarke, p.85.
13. Goodrick-Clarke, *The Occult Roots of Nazism*, p.63.
14. Hitler, op. cit., p.244.
15. Matthew and Harrison, op. cit.
16. Chamberlain, *Foundations of the 19th Century*, p.xxxvii.

17. Chamberlain, *Immanuel Kant, A Study and a Comparison with Goethe, Leonardo da Vinci, Bruno, Plato and Descartes*, p.50.
18. Ibid, p.107.
19. Ibid, p.118.
20. Ibid, p.118.
21. Ibid, p.123.

CHAPTER 37

1. Oxford Dictionaries Online.
2. Ibid.
3. Dawkins presents 'The Genius of Charles Darwin', C. IWC Media Ltd, Channel 4 DVD, 2008.
4. Dawkins, *The Selfish Gene*, p.192.
5. Oxford Dictionaries Online.
6. Ibid.
7. Dawkins, op. cit., p.192.
8. Ibid, p.323.
9. Ibid, pp.191–3.
10. Dawkins, op. cit., p.200.
11. Ibid, Endnotes, p.323.

CHAPTER 38

1. Dawkins presents 'The Genius of Charles Darwin', op. ct.
2. Oxford Dictionaries Online.
3. Ibid.
4. Litchfield, *Emma Darwin*
5. Oxford Dictionaries Online.
6. Baron-Cohen, Simon, *Zero Degrees of Empathy*, p.87.
7. Oxford Dictionaries Online.
8. Newberg and Walman, *How God Changes Your Brain*, pp.55–6.

CHAPTER 39

1. Oxford Dictionaries Online.
2. Spector, *Identically Different: Why you can Change your Genes*, p.11.
3. Ibid.
4. Cubas, Vincent and Coen. 'An Epigenetic Mutation Responsible for Natural Variation in Floral Symmetry'. *Nature*, 401: pp.157–61, in Spector, op. cit., pp.34–6.
5. Spector, op cit., p.36.
6. Oxford Dictionaries Online.
7. Ibid.
8. Spector, op. cit., p.65.
9. Ibid, p.107.
10. Ibid, p.148.
11. Ibid, p.151.

NOTES

12. Ibid, p.163.
13. Ibid, p.197.
14. Ibid, p.259.
15. Ibid, p.164.
16. Ibid, p.293.
17. Ibid, pp.39–40.
18. Ibid, p.293.

CHAPTER 40

1. Clarke and Claydon, 'Darwin's Church', by Paul White, *Studies in Church History*, Volume 46, p.344.
2. Oxford Dictionaries Online.
3. Clarke and Claydon, op. cit., p.345.
4. Ibid, p.346.
5. Innes to Darwin, 21 January 1871, Cor.19, p.29.
6. Darwin to Innes, 15 December, Darwin Correspondence Project, Letter 3343.
7. Darwin to Innes, 1 May, Darwin Correspondence Project, Letter 3528.
8. Oxford Dictionaries Online.
9. Horsman to Darwin, 2 June, Darwin Correspondence Project, Letter 6223, and Clarke and Claydon, op. cit., pp.346–7.
10. Darwin to Innes, 20 January, Cor.16, p.28.
11. Darwin to Innes, 15 June, Darwin Correspondence Project, Letter 6242.
12. Darwin to Innes, 1 December 1868, Cor.16, pp.871-2.
13. Darwin to Innes, 10 December, Darwin Correspondence Project, Letter 6497.
14. Darwin to Innes, I December 1868, Darwin Correspondence Project, Letter 6486, and Clarke, and Claydon, op. cit., p.348.
15. Darwin to Innes, 18 January, Darwin Correspondence Project, Letter 7445.
16. Emma to F.P. Cobbe, [25 February 1871], Cor.19, p.106.
17. Ffinden, to Emma, 24 December 1873, cited in Moore, op. cit., p.471, and Clarke and Claydon, op. cit., Volume 46, pp.348–9.
18. Ffinden to Lubbock, 1875, cited in Moore, op. cit., pp.471-2, and Clarke and Claydon, Op cit., Volume 46, pp.348–9.
19. Clarke and Claydon, op. cit., Volume 46, pp.348–9.
20. Darwin to Ffinden, 21 May, Darwin Correspondence Project, Letter 8342.
21. Darwin to Ffinden, G.S., Darwin Correspondence Project, Letter 10706a.
22. Darwin to Innes, 27 November, Darwin Correspondence Project, Letter 11763.
23. Darwin to Henry N. Ridley, 28 November 1878, Darwin Correspondence Project, Letter 11766.
24. Moore, op. cit., p.473.
25. Darwin to Fegan, Darwin Correspondence Project, Letter 12879.
26. Clarke and Claydon, op. cit., Volume 46, p.350.

CHAPTER 41

1. Oxford Dictionaries Online.

2. Cor.4, p.411.
3. Cor.4, p.412.
4. Cor.4, p.415.
5. Cor.4, p.424.
6. Darwin to W. E. Darwin, 3 October, Cor.5, p.63.
7. Darwin to W.E. Darwin, 24, Cor.5, p.81.
8. Darwin to Fox, 17 July, Cor.4, p.148.
9. Keynes, *Annie's Box: Charles Darwin, His Daughter and Human Evolution,* pp.200–1. 10. Steensma, 'Down syndrome in Down House: trisomy 21, *GATA1*mutations, and Charles Darwin', *Blood: Journal of the American Society of Hematology*, Volume 105, No.6, pp.2614-16
11. Darwin to William Darwin Fox, 27 March 1851, Cor.5, p.9.
12. Darwin to Emma, 18 April 1851, Cor.5, p.14.
13. Cor.5, p.14, note 2.
14. Darwin to Fox, 29 April 1851, Cor.5, p.32.
15.Darwin to Jenyns, 1 April, Darwin Correspondence Project, Letter 2251.
16. Oxford Dictionaries Online.
17. Darwin to Lubbock, 17 July 1870, Cor.18, p.215.
18. Darwin, *The Descent of Man,* p.688.
19. Berra, Alvarez and Ceballos, 'Was the Darwin/Wedgwood Dynasty Adversely Affected by Consanguinity?', In *BioScience*, May 2010/Vol.60, No.5, pp.376–83.
20. Oxford Dictionaries Online.
21 Carvalho, Luiz O.P., and others. 'Trypanoma cruzi and myoid cells from seminiferous tubules: interaction and relation with fibrous components of extracellular matrix in experimental Chagas' disease. *International Journal of Experimental Pathology*, 2009, February: 90(1): pp.52-7.

CHAPTER 42

1. Emma to Fanny Allen, 21 January 1873, in Litchfield, Henrietta, op. cit., p.211.
2. Matthew and Harrison, op. cit.
3. Oxford Dictionaries Online.
4. Conway, *Centenary History of the South Place Society*, p.19.
5. Ibid, p.55.
6. Ibid, p.58.
7. Ibid, p.58.
8. 'The Story of the South Place Ethical Society' by Norman Bacrac. Based on a talk given to the Farnham Humanists, 24 February 2008.
9. Conway, Moncure Daniel, *Centenary History of the South Place Society*, p.53.
10. Ibid, p.53
11. Ibid, p.79.
12. Ibid, p.83.
13. Ibid, p.86.
14. Darwin to Nicolaas D. Doedes, 2 April 1873, Darwin Correspondence Project, Letter 8837.
15. Darwin to Lyell, 3 September, Darwin Correspondence Project, Letter 9621.
16. Darwin to Grant, 11 March 1878, Darwin Correspondence Project, Letter 11416.
17. Darwin to Ridley, 28 November 1878, Darwin Correspondence Project, Letter 11766.
18. Emma to Mengden, 8 April 1879, Darwin Correspondence Project, Letter 11981.

19. Darwin to Fordyce, 7 May 1879, Darwin Correspondence Project, Letter 12041.
20. Oxford Dictionaries Online.
21. Darwin to McDermott, 24 November 1880, Darwin Correspondence Project, Letter 12851.
22. Oxford Dictionaries Online.
23. *The Academy*, 10 June 1882, No.527, p.417.
24. Oxford Dictionaries Online.
25. Howarth and Howarth *A History of Darwin's Parish,*
26. Wallace, op. cit., pp.237–8.
27. Conway, op. cit., p.49.
28. Ibid, p.49
29. Litchfield, op. cit., p.260.
30. Ibid, p.270.

EPILOGUE

1. 'Beautiful Minds', Episode 3, Series 2. BBC4, 2012.
2. Attenborough, 'This Heaving Planet', *New Statesman*, 27 April 2011.
3. 'Darwin's Struggle: the Evolution of The Origin of Species', BBC Productions, Bristol, 2009.
4. Dawkins presents 'The Genius of Charles Darwin', 'The Enemies of Reason', and 'Root of all Evil?', op. cit.
5. www.nhm.ac.uk/about-us/news/2008/may/darwins-statue-on-the-move13846.html

APPENDIX I

1. FitzRoy, *A Narrative of the Voyage of* HMS *Beagle*, p.20.

Bibliography

Adler, Professor Saul, 'Darwin's Illness'. *Nature*, Vol. 184, pp.1102–03, 10 October 1959

American Psychiatric Association, *Diagnostic and Statistical Manual of Mental Disorders,* (DSM-IV-TR) (American Psychiatric Association, Washington, DC, 2000)

Arendt, Hannah, *The Origins of Totalitarianism* (Harcourt, New York, 1966)

Barlow, Emma (editor), *The Autobiography of Charles Darwin, 1809–1882* (with original omissions restored) (Collins, London, 1958)

Baron-Cohen, Simon, *Zero Degrees of Empathy* (Allen Lane, London, 2011)

Bateson, Dusha and Weslie Janeway, *Mrs Charles Darwin's Recipe Book* (Glitterati, New York, 2008)

Berg, Jeremy M., John L. Tymoczko, Lubert Stryer, and Neil D. Clarke, *Biochemistry* (W. H. Freeman, New York, 1995)

Boon, Nicholas A., Nicki R. Colledge, Brian R. Walker, and John A. A. Hunter (editors), *Davidson's Principles and Practice of Medicine* (Churchill Livingstone Elsevier, London and New York, 2006)

Brochu, Christopher A., et al., *Dinosaurs: the Ultimate Guide to Prehistoric Life* (HarperCollins, London, 2000)

Burkhardt, Frederick and others (editors), *The Correspondence of Charles Darwin:* in nineteen volumes (Cambridge University Press, Cambridge, 1985–2012)

Calman, William T., 'Patrick Matthew of Gourdiehill, Naturalist' (1912). Published in the British Association's *Handbook and Guide to Dundee and District* (A. W. Paton and A. H. Millar, editors)

Chamberlain, H. S., *Foundations of the 19th Century*. (Second Edition) (John Lane, The Bodley Head, London, 1912)

Chamberlain, Houston Stewart, *Immanuel Kant: A Study and a Comparison with Goethe, Leonardo da Vinci, Bruno, Plato and Descartes*. (Authorized translation from the German by Lord Redesdale, with an introduction by the translator.) (John Lane, The Bodley Head, London, 1914)

Chambers, Robert, *Vestiges of the Natural History of Creation* (published anonymously in 1884). (The Echo Library, Teddington, Middlesex)

Chase, Allan, *The Legacy of Malthus: The Social Costs of the New Scientific Racism* (University of Illinois Press, Champaign, Illinois, 1980)

Clarke, Peter and Tony Claydon (editors), 'Darwin's Church', by Paul White. *Studies in Church History*, Volume 46 (Ecclesiastical History Society, 2010)

Colp, Ralph (Jr), *Darwin's Illness* (University Press of Florida, Gainesville, Florida, 2008)

Conway, Moncure Daniel, *Centenary History of the South Place Society: Based on Four Discourses given in the Chapel in May and June, 1893* (General Books, Memphis, TN, 2009)

Cox, Professor F. E. G., *The Wellcome Trust Illustrated History of Tropical Diseases* (The Wellcome Trust, London, 1996)

Cresswell, Beatrix F., *Exeter Churches* (James G. Commin, Exeter, 1908)

Daim, Dr Wilfried, *Der Mann, der Hitler die Ideen gab.* 2nd edn. (Isar Verlag, Vienna/Cologne/Graz, 1985)

, *Beagle Diary*, http://darwinonline.org.uk

, *The Descent of Man and Selection in Relation to Sex* (Penguin Books, London, 2004)

, *The Expression of the Emotions in Man and Animals* (first published in 1872). (HarperCollins, London, 1998)

, (First published 1879). *The Life of Erasmus Darwin* (Cambridge University Press, Cambridge, 2003)

, *The Origin of Species by Means of Natural Selection or the Preservation of Favoured Races in the Struggle for Life*. (First published in 1859) (The New American Library, New York and Toronto, and The New English Library, London, 1958)

, *The Variation of Animals and Plants under Domestication*. Volume I. (First published in 1868). (Echo Library, Teddington, Middlesex, 2007)

, *The Variation of Animals and Plants under Domestication*. Volume II (John Murray, London, 1868)

, *The Voyage of the Beagle*. (First published 1845) (White Star, Vercelli, Italy, 2006)

, *Geological observations on the volcanic islands visited during the voyage of* HMS *Beagle, together with some brief notices of the geology of Australia and the Cape of Good Hope. Being the second part of the geology of the voyage of the Beagle, under the command of Capt FitzRoy, RN during the years 1832 to 1836* (Smith Elder, London, 1844)

Darwin, Erasmus, *Zoonomia; or The Laws of Organic Life*, Vol. I. LONDON: Printed for J. Johnson, in St Paul's

BIBLIOGRAPHY

Churchyard, 1796 (A Project Gutenberg e-book, release date: April 25, 2005.)
, *Zoonomia; or The Laws of Organic Life*, Volume II (J. Johnson, London, 1796)
Darwin, Francis (editor), *Autobiography of Charles Darwin* (Icon Books, Cambridge, 2003)
(editor), *The Life and Letters of Charles Darwin, Including an Autobiographical Chapter* (Bibliobazaar, Charleston, SC, 2006)
(editor), *The Life of Charles Darwin* (re-edited by J.P. de Boulogny, Lulu.com, 2009)
, and A. C. Seward (editors), *More Letters of Charles Darwin: a record of his work in a series of hitherto unpublished letters*. 2 vols. (John Murray, London, 1903)
Dawkins, Richard, *The Selfish Gene* (Oxford University Press, Oxford, 2006)
Dickens, Margaret, *A Thousand Years in Tardebigge* (Cornish Brothers, Birmingham, 1931)
Edwards, Clive, *Encyclopedia of Furnishing Textiles, Floorcoverings and Home Furnishing Practices, 1200–1950* (Lund Humphries Hampshire, and Burlington, VT, 2007)
FitzRoy, Captain Robert, RN, *Narrative of the Surveying Voyages of His Majesty's Ships* Adventure *and* Beagle *between the years 1826 and 1836* (Henry Colburn, London, 1839)
Flew, Antony (editor), *Thomas Robert Malthus, An Essay on the Principle of Population* (first published London, 1798), and *A Summary view of the Principle of Population* (first published 1830). (Penguin Books, London, 1970)
Forbes, Edward, Essay 'On the Connexion Between the Distribution of the Existing Fauna and Flora of the British Isles', published in *The Memoirs of the Geological Survey of Great Britain and of the Museum of Economic Geology in London*, vol.1 (Longman, Brown, Green and Longmans, London, 1846)
Gelder, Michael, Paul Harrison, and Philip Cowen, *Shorter Oxford Textbook of Psychiatry* (Oxford University Press, Oxford, 2006)
Galton, Francis, *Hereditary Genius: An Enquiry into its Laws and Consequences*. (First published in 1869). (LLC Publication, General Books, New York, 2009)
, *Inquiries into Human Faculty and its Development* (J. M. Dent & Sons, London, 1928)
Goodrick-Clarke, Nicholas, *The Occult Roots of Nazism* (New York University Press, NY, 1992)
Gully, James Manby, MD, *The Water-cure in Chronic Diseases* (S. R. Wells, New York, 1880)
Hamann, Brigitte, *Hitler's Vienna: a Dictator's Apprenticeship* (Oxford University Press, Oxford, 1999)
Hitler, Adolf, *Mein Kampf* (Jaico Publishing House, Mumbai, 2009)
Huxley, Julian, *What Dare I Think? The Challenge of Modern Science to Human Action & Belief* (Chatto & Windus, London, 1931)
Keynes, Randal H., *Annie's Box: Charles Darwin, His Daughter and Human Evolution* (Fourth Estate, London, 2001)
Kliegman, Robert M., Richard E. Behrman, Hal B. Jenson and Bonita F. Stanton, *Nelson Textbook of Pediatrics* (Saunders Elsevier, Philadelphia, PA, 2007)
Lamarck, J.B., *Zoological Philosophy* (Hafner Publishing, New York and London, 1963)
List, Guido von, *The Secret of The Runes* (Edited, introduced, and translated by Stephen E. Flowers). (Destiny Books, Vermont, 1988)
Litchfield, Henrietta, *Emma Darwin* (Privately printed edition). (Cambridge University Press, Cambridge, 1904)
(editor), *Emma Darwin: A Century of Family Letters 1792–1896*, Volumes I and II (D. Appleton, New York, 1915)
Locke, John, *An Essay Concerning Human Understanding* (London, 1690)
Loudon, J. C. (conductor), *Gardeners' Magazine and Register of Rural and Domestic Improvement* (Longman, Rees, Orme, Brown, and Green, London)
Mandell, Gerald L., John E. Bennett and Raphael Dolin (editors) *Principles and Practice of Infectious Diseases* (Churchill Livingstone, London and New York, 2000)
Marshall, Barry, 13 February 2009. 'What I Know and What I Think I Know'. http://barryjmarshall.blogspot.com/2009/02/darwins-illness-was helicobacter-pylori.html
Martineau, James, *A Seat of Authority in Religion*, in *James Martineau: Selections*, edited by Alfred Hall. 1950 (Lindsey Press, London, 1890)
Matthew, H. C. G., and Brian Harrison (editors) *Oxford Dictionary of National Biography* (Oxford University Press, Oxford, 2004)
Matthew, Patrick, *On Naval Timber and Arboriculture; with Critical Notes on Authors who have recently treated the Subject of Planting* (Longman, Rees, Orme, Brown, and Green, London, and Adam Black, Edinburgh, 1831)
Moore, James, 'Darwin of Down: the Evolutionist as Squarson-naturalist', in *The Darwinian Heritage*, edited by David Kohn (Princeton University Press in association with Nova Pacifica, Princeton, NJ, 1985)

Morgan, Kenneth O., *The Oxford Illustrated History of Britain* (Guild Publishing, London, 1986)
Murphy, Martin, *Blanco White: Self-banished Spaniard* (Yale University Press, New Haven and London, 1989)
Newberg, Andrew, and Mark R. Walman, *How God Changes Your Brain* (Ballantine Books, New York, 2009)
Nietzsche, Friedrich, *Also sprach Zarathustra: Ein Buch für alle und keinen, IV Die Begrussung* (Alfred Krõner, Stuttgart, 1926)
Nietzsche, Friedrich, *The Will to Power in Science, Nature, Society and Art* (Random House, New York, 1968)
Norman, Andrew T. P., *Hitler: Dictator or Puppet* (Pen & Sword Books, Barnsley, 2011)
Oldham, J Basil, *A History of Shrewsbury School 1552–1952* (Basil Blackwell. Oxford, 1952)
Orel, Vitezslav, *Heredity before Mendel* (Oxford University Press, 1996)
Paley, Archdeacon William, *A View of the Evidences of Christianity* (R. Faulder, London, 1804)
Paley, W., *Natural Theology, or Evidences of the Existence and Attributes of the Deity, Collected from the Appearances of Nature*; Bridgewater Treatises (R. Faulder, London, 1803, reissued by Cambridge University Press, Cambridge, 2009)
Paley, William, *A View of the Evidences of Christianity* (R. Faulder, London, 1746)
Pearson, Karl, *National Life from the Standpoint of Science* (second edition). (Adam & Charles Black, London, 1919)
Paton, A. W and A. H. Millar (editors), British Association, Dundee Meeting, *Handbook and Guide to Dundee and District* (David Winter & Son, Dundee, 1912)
Prange, Gordon W., (ed.). *Hitler's Words* (American Council on Public Affairs, 1944)
Price, Gregory D. and Elizabeth V Nunn, 'Valanginian Isotope Variation in Glendonites and Belemonites from Arctic Svalvard: Transient Glacial Temperatures during the Cretaceous Greenhouse'. *Geology*, March 2010: 38; pp.251–4
Quarterly Journal of Agriculture (William Blackwood, Edinburgh and T. Cadell, Strand, London)
Rauschning, Hermann, *Hitler Speaks* (Thornton Butterworth, London, 1939)
Reed, The Reverend Clifford M., *Beatrix Potter's Unitarian Context*, published in *Beatrix Potter: Thirty Years of Discovery and Appreciation*, Joy & Judy Taylor (editors). (Beatrix Potter Society, 2010)
Reed, The Reverend Clifford M., *Unitarian? What's that?* (Lindsey Press, London, 1999)
Reeve, Tori, *Down House: the Home of Charles Darwin* (English Heritage, Holburn, London, 2009)
Rose, Steven, *The Making of Memory* (Vintage, London, 2003)
Royal Society, The: 350 years of Science. Exhibition catalogue, issued July 2010
Shirer, William L., *The Rise and Fall of the Third Reich* (Book Club Associates, London, by arrangement with Secker and Warburg Limited, 1972)
Simpson, J. A. and E. S. C. Weiner (prepared by), *The Oxford English Dictionary* (Clarendon Press, Oxford, Oxford, 1989)
Spector, Tim, *Identically Different: Why you can Change your Genes* (Weidenfeld & Nicolson, London, 2012)
Stanbury, David (editor), *A Narrative of the Voyage of* HMS *Beagle* (The Folio Society, London, 1977)
Thompson, Della, *The Concise Oxford Dictionary* (BCA, London, 1998)
Tritton, Roger (managing editor), *The Hutchinson Encyclopedia for the Millennium* (Best Sellers Direct, Sheffield, UK, 2000)
United Services Journal and Naval and Military Magazine, 1831 Part II, pp.457–66, and 1831 Part III, pp.65–76 (Henry Colburn and Richard Bentley, London, kindly supplied by the British Library)
Wallace, Alfred R, *My Life: A Record of Events and Opinions* (Chapman & Hall, London, 1908)
Wallace, Alfred Russel, (first published in 1905). *My Life: A Record of Events and Opinions* (Elibron Classics, New York, 2005)
Warrell, David A., Timothy M. Cox, John D. Firth, and Edward J. Benz Jr (editors) *Oxford Textbook of Medicine* (Oxford University Press, Oxford, 2003)
Wollaston, T. Vernon, *On the Variation of Species, with Special Reference to the Insecta; followed by an Enquiry into the Nature of Genera* (John Van Voorst, London, 1856)
Wells W. C., 'Two Essays: One Upon Single Vision with Two Eyes; the Other on Dew' (Constable, London, 1818)

Film Documentaries

'The Nazis: A Warning from History: the Wrong War'. A BBC/A&E Network Production. C. BBC MCMXCVII.
'Beautiful Minds', Episode 3, Series 2, BBC4, 2012.
Richard Dawkins presents 'The Genius of Charles Darwin', 'The Enemies of Reason', and 'Root of all Evil?', c. IWC Media Ltd, Channel 4 DVD, 2007, 2008.

Index

INDEX

INDEX

INDEX